LONDON MATHEMATICAL SOCIETY LECTURE NOTE SERIES

Managing Editor: Professor J.W.S. Cassels, Department of Pure Math[...]
University of Cambridge, 16 Mill Lane, Cambridge CB2 1SB, Englan[...]

The titles below are available from booksellers, or, in case of difficult[...]

London Mathematical Society Lecture Note Series. 220

Algebraic Set Theory

A. Joyal
Université du Québec à Montréal

I. Moerdijk
Universiteit Utrecht

CAMBRIDGE
UNIVERSITY PRESS

Published by the Press Syndicate of the University of Cambridge
The Pitt Building, Trumpington Street, Cambridge CB2 1RP
40 West 20th Street, New York, NY 10011-4211, USA
10 Stamford Road, Oakleigh, Melbourne 3166, Australia

© Cambridge University Press 1995

First published 1995

Printed in Great Britain at the University Press, Cambridge

Library of Congress cataloging in publication data

Joyal, André.
Algebraic set theory / André Joyal and Ieke Moerdijk.
    p.   cm. - (London Mathematical Society lecture note series; 220)
Includes bibliographical references (p.   -   ) and index.
ISBN 0-521-55830-1 (pbk.)
1. Set theory. I. Moerdijk, Ieke. II. Title. III. Series.
QA248.J69 1995
511.3'22--dc20   95-15173   CIP

British Library cataloguing in publication data available

ISBN   0 521 55830 1 paperback

Contents

Preface

This small book grew out of a desire to perceive the hierarchy of sets as an algebraic structure of a simple kind. We began our work on the hierarchy and on small maps around 1988, simultaneously with related work on open maps (see our (1990), (1994) in the bibliography). The main results were announced in our (1991), and presented in several lectures between 1990 and 1993.

We are most grateful to J. Bénabou, M. Jidbladze and J. van Oosten for their valuable comments on earlier drafts of this book, and to Elise Goeree for the careful typing of the manuscript.

Our work would not have been possible without the support for many mutual visits, by the Canadian National Science and Engineering Research Council and the Netherlands Organization for Scientific Research (NWO).

Montreal and Utrecht

A. Joyal
Département de Mathématiques
Université du Québec à Montréal
Montréal, H3C 3P8
Canada

I. Moerdijk
Mathematisch Instituut
Universiteit Utrecht
3508 TA   Utrecht
Nederland

Introduction

In this book we present a formalization of set theory based on operations on sets, rather than on properties of the membership relation. The two operations are union and successor (singleton), and the algebras for these operations will be called Zermelo-Fraenkel algebras. The definition of these algebras uses an abstract notion of "small map". We show that the usual axioms of Zermelo-Fraenkel set theory are nothing but a description of the free ZF-algebra, just as the axioms of Peano arithmetic form a description of the free monoid on one generator.

The basic ideas are quite simple, and could roughly be explained as follows. Imagine a "universe of sets" $\mathcal{C}$, in which one distinguishes some sets as "small". For example, one could take for $\mathcal{C}$ all the countable sets, and call a set small if it is finite. Another example is provided by taking for $\mathcal{C}$ all the classes (in the sense of set theory), and calling a class small if it is a set, rather than a proper class. In such a universe $\mathcal{C}$ we consider partially ordered sets L which have the property that each "small" subset of L has a supremum, and which are equipped with a distinguished operation $s : L \to L$, called successor. Thus L could be thought of as an algebraic structure, with rather a lot of operations: besides the successor s, which is a unary operation, there is for each small set $S \in \mathcal{C}$ an S-ary operation on L, given by the S-indexed supremum.

In spite of this multitude of operations, it is possible to apply many constructions and results from algebra to such "algebras" L. In fact, they are not so different from, for example, the standard differential algebras: our algebras have small sups instead of finite sums, and a successor s instead of a differential d; they do not satisfy the identity $d \circ d = 0$ of differential algebra, but we shall consider other identities for the successor s.

For two algebras L and L', a homomorphism from L to L' is of course a mapping $\varphi : L \to L'$ which commutes with the operations. In other words, φ preserves small suprema, and commutes with the successor. With these homomorphisms, one can apply the usual constructions of algebra by "generators and relations". For example, the free algebra on a set A, which we

1

denote by $V(A)$, is uniquely defined by the property that for any algebra L and any mapping $A \to L$, there is a uniquely defined extension to a homomorphism $V(A) \to L$. The relation to formal (Zermelo-Fraenkel) set theory now becomes apparent: in the example, mentioned above, where $\mathcal{C}$ consists of all the classes, $V(A)$ is essentially the cumulative hierarchy of sets built on A as a collection of atoms. In the example above where $\mathcal{C}$ consists of countable sets and "small" means finite, the free algebra $V(\emptyset)$ on the empty set is the algebra of hereditarily finite sets. A typical example of adding a "relation" is the algebra O, freely generated by the condition that the successor is monotone. In the example where $\mathcal{C}$ consists of all the classes, O is the class of ordinal numbers. In the example where $\mathcal{C}$ consists of countable sets, O is the set of natural numbers. The algebraic properties of O capture transfinite induction for ordinals, and ordinary induction for natural numbers, respectively.

Of course, the theory can only be developed when one assumes that the collection of "small" sets satisfies some suitable axioms. To get off the ground at all, we will assume that the empty set and the one-point set are both small. Thus, in particular, any algebra L contains the supremum of the empty subset of L; i.e. L must have a smallest element 0. Furthermore, we will assume that the union of a small family of small sets is small, and that the disjoint sum of two small sets is small. Thus, in particular, the two-point set is small, and hence any algebra L has an operation $\vee : L \times L \to L$ of binary supremum. There are also axioms for covers (i.e., surjections): if $S \twoheadrightarrow T$ is a cover and S is small then T is small; and conversely, if T is small then S contains a small subset $S' \subseteq S$ which already covers T. In the example where $\mathcal{C}$ consists of all classes, this latter property for the existence of "small subcovers" is usually referred to as the *collection axiom* of set theory. Finally, the following two axioms play a crucial rôle in the construction of algebras by generators and relations. First, for any set C in $\mathcal{C}$ and any small set S in $\mathcal{C}$, one can form the set (not necessarily small) C^S of all functions from S to C. Secondly, there exists a "universal" small set in $\mathcal{C}$: this is a mapping $\pi : E \to U$ in $\mathcal{C}$ such that for any small set S there is some point $x \in U$ for which S is isomorphic to $\pi^{-1}(x)$. These are essentially all the axioms that we will ask the collection of "small" sets in the universe $\mathcal{C}$ to satisfy.

It is important to observe that it is only necessary for the "universe of sets" $\mathcal{C}$ in which we construct and study the algebras L to possess some very basic properties. It should be possible to interpret the basic operations of the first order logic (conjunction, disjunction, universal and existential quantification, etc.) in $\mathcal{C}$. This means, for example, that for two mappings $\alpha : S \to T$ and $\beta : R \to S$ in $\mathcal{C}$ one can form the set $\{t \in T \mid \forall s \in S :$

if $\alpha(s) = t$ then $\exists\ r \in R\ :\ \beta(r) = s\}$ inside the universe $\mathcal{C}$. These logical operations need not even satisfy the rules of classical first order logic; but they should satisfy at least the rules of intuitionistic logic. There are many interesting examples of such universes $\mathcal{C}$ which are quite different from the usual universe of sets. Thus, $\mathcal{C}$ can be a universe of *sheaves* (i.e., sets which vary continuously over a fixed topological space of parameters), or an "effective" universe of *recursive* sets. More generally, $\mathcal{C}$ can be any *elementary topos* (see Mac Lane-Moerdijk (1992) for this notion and many examples of elementary topoi). In this way, our theory extends both topos theory and (intuitionistic) set theory. In particular, the theory is powerful enough to capture in a constructive way the theory of ordinals and of transfinite induction.

In fact, one of our main motivations was the apparent discrepancy between sheaves and topoi on the one hand, and models of Zermelo-Fraenkel set theory on the other. Elementary topoi correspond naturally to a weak kind of set theory with only bounded quantifiers, as discussed extensively in the early topos literature (cf. Mitchell(1972), Cole(1973), Osius(1974), and others). For an arbitrary topos, it is in general not possible to build a corresponding model for Zermelo-Fraenkel set theory. Nevertheless for many topoi which are constructed using set theory to begin with (such as topoi of Boolean sets or of sheaves), one can obtain correponding models for Zermelo-Fraenkel set theory by a transfinite iteration of the power-set operation of the topos along the "external", classical, ordinal numbers. For Boolean sets, this construction goes back to Scott and Solovay (see Bell(1977)). For general sheaves, it is discussed in Fourman(1980), Freyd(1980), Blass-Scedrov(1989), and elsewhere.

To describe the variety of different examples of universes $\mathcal{C}$ having the required properties, and to exploit relations with topos theory, we will formulate our theory using the language of categories (Mac Lane(1971)). For readers who are not sufficiently familiar with this language, we hasten to point out that much of this book can be read and understood at a less general level, by assuming throughout that $\mathcal{C}$ is an actual universe of sets, as in the two examples – countable sets and classes – mentioned above.

In the language of category theory, the "small" objects in $\mathcal{C}$ will (have to) be described in terms of small maps. Intuitively, these are the maps $f : E \to X$ all of whose fibers $f^{-1}(x)$ are small; a small map $E \to X$ in $\mathcal{C}$ can be thought of as a continuous family of small objects, parametrized by X. A basic axiom for these small maps, which expresses that smallness is a property of the fibers of the map, is that in a pullback square

the map $E' \to X'$ is small whenever $E \to X$ is, while conversely, if $E' \to X'$ is small and $X' \to X$ is surjective then $E \to X$ is small. Other axioms are direct translations of the axioms for small sets mentioned before. For example, the axiom that a small union of small sets is small can now simply be expressed by stating that the composition of two small maps is again small.

The idea of continuous families of "small" objects, constructed as mappings $E \to X$ with suitable properties, is ubiquitous in geometry and physics. Well-known examples include live bundles (families of lines) and proper maps (families of compact spaces) in topology, and families of curves in algebraic geometry. In this context, one often studies universal families of such small objects, such as classifying spaces for line bundles (projective spaces and Grassmann manifolds) and moduli spaces of curves. When stated for small maps, our axiom for a universal small set takes a similar form: it states that there is an object U with a small map $\pi : E \to U$, such that every small map is locally a pullback of this universal small map $\pi : E \to U$. Thus, U is a "classifying space" for small maps, having properties much like classifying spaces for vector bundles and other well-known classifying spaces in topology. (To illustrate the analogy, we explain at the end of Chapter I how the classifying space U for small maps is "unique up to homotopy".)

Our abstract framework thus consists of a suitable category $\mathcal{C}$, with a designated class of arrows in $\mathcal{C}$, which are called small, and satisfy natural axioms. In this general context, it is possible to define algebras L as objects in $\mathcal{C}$ equipped with an operation $s : L \to L$ for successor, and with a partial order on L which is complete in the sense that the supremum exists along any map which is designated as small. Such algebras L will be called *Zermelo-Fraenkel algebras* in $\mathcal{C}$. We investigate the structure of the free (initial) ZF-algebra V, and show that it can be viewed as an algebra of small sets, via an explicit isomorphism between V and the object $P_s(V)$ of "small subsets" of V. This free algebra V should be viewed as the cumulative hierarchy of small sets, relative to the ambient category $\mathcal{C}$ and its class of small maps. Indeed, we prove in Chapter II, §5, that under very general conditions this algebra V is a model of the axioms for (intuitionistic) Zermelo-Fraenkel set theory. In Chapter IV, we will explain in detail how one obtains, as particular examples, the sheaf models and effective (realizability) models for set theory already referred to above.

Our algebraic approach also makes it possible to distinguish different types of ordinal numbers in a very natural way. For example, in Chapter II, §2, we will discuss how, within the category $\mathcal{C}$, the ZF-algebra O which is free on a monotone successor operation $t : O \to O$ enables one to write V as a cumulative hierarchy of objects V_α, suitably indexed by elements $\alpha \in O$. The classical Von Neumann ordinals, defined as hereditarily transitive sets, also appear as a free algebra, generated by the relation that the successor is inflationary ($x \leq s(x)$). Furthermore, in Chapter II, §4, it will be discussed how the ZF-algebra T which is free on a successor $r : T \to T$ preserving binary suprema enables one to give a purely constructive proof of Tarski's fixed point theorem, using "transfinite induction" along this object T.

In Chapter III, it will be shown how all these free algebras can be explicitly constructed as objects of the ambient category $\mathcal{C}$. This construction of free algebras makes use of the theory of open maps, and of (bi-)simulations for trees and forests. The explicit use of bisimulation for set theory goes back to the work on non-well-founded sets by Aczel(1988). It would be of interest to construct sheaf models for the theory of non-well-founded sets from our axioms for small maps.

Chapter I

Axiomatic Theory of Small Maps

§1   Axioms for small maps

In this first section we will present a set of axioms for a class $\mathcal{S}$ of small maps in a category $\mathcal{C}$. These axioms are meant to express some basic properties of maps with "small" fibers. For example, if $\mathcal{C}$ is the category of sets, our axioms are satisfied by the class of maps with finite fibers (those $f : Y \to X$ with $f^{-1}(x)$ finite for each $x \in X$), or the class of maps with countable fibers, etc.

The ambient category $\mathcal{C}$ will be assumed to be a Heyting pretopos with a natural numbers object. This means that $\mathcal{C}$ is a category with enough structure to interpret (intuitionistic) first order logic and arithmetic. (We recall the precise definition in Appendix B.)

Our axioms for small maps are an extension of the axioms for open maps presented in Joyal-Moerdijk(1990) and (1994), and we begin by recalling those. Consider the following properties of a class $\mathcal{S}$ of arrows in the category $\mathcal{C}$.

(**A1**) Any isomorphism belongs to $\mathcal{S}$, and $\mathcal{S}$ is closed under composition.

(**A2**) ("Stability") In any pullback square

$$
\begin{array}{ccc}
Y' & \longrightarrow & Y \\
{\scriptstyle g}\downarrow & & \downarrow{\scriptstyle f} \\
X' & \xrightarrow{\;p\;} & X
\end{array}
\tag{1}
$$

7

if f belongs to $\mathcal{S}$ then so does g.

(A3) ("Descent") In any pullback square (1), if g belongs to $\mathcal{S}$ and p is epi then f belongs to $\mathcal{S}$.

(A4) The maps $0 \to 1$ and $1 + 1 \to 1$ belong to $\mathcal{S}$.

(A5) ("Sums") If $Y \to X$ and $Y' \to X'$ belong to $\mathcal{S}$ then so does their sum $Y + Y' \to X + X'$.

(A6) ("Quotients") In any commutative diagram

$$
\begin{array}{ccc}
Z & \xrightarrow{\;\;p\;\;} & Y \\
& \searrow{\scriptstyle g} \quad \swarrow{\scriptstyle f} & \\
& B &
\end{array}
\tag{2}
$$

if p is epi and g belongs to $\mathcal{S}$ then so does f.

(A7) ("Collection Axiom") For any two arrows $p : Y \twoheadrightarrow X$ and $f : X \to A$ where p is epi and f belongs to $\mathcal{S}$, there exists a quasi-pullback square of the form

$$
\begin{array}{ccc}
Z \longrightarrow Y & \xrightarrow{\;p\;} & X \\
\downarrow{\scriptstyle g} & & \downarrow{\scriptstyle f} \\
B \xrightarrow[\;h\;]{} & & A
\end{array}
\tag{3}
$$

where h is epi and g belongs to $\mathcal{S}$.

(Recall that such a square is said to be a quasi-pullback if the obvious arrow $Z \to B \times_A X$ is an epimorphism.)

The class $\mathcal{S}$ is said to be a class of *open maps* (with collection) if it satisfies these axioms (A1 − 7). For standard examples of such classes we refer the reader to Joyal-Moerdijk(1994).

Before we state our axioms for a class of small maps, we recall that a map $f : Y \to X$ in $\mathcal{C}$ is said to be *exponentiable* if f is exponentiable as an object of the slice category $\mathcal{C}/X$ (i.e., the functor $f^* : \mathcal{C}/X \to \mathcal{C}/X$ sending $Z \to X$ to $Z \times_X Y \to X$ has a right adjoint).

1.1 Definition. A class $\mathcal{S}$ of arrows in a category $\mathcal{C}$ is said to be a *class of small maps* if $\mathcal{S}$ is a class of open maps (with collection) satisfying the following two additional axioms (S1) and (S2).

(S1) ("Exponentiability Axiom") Every map in $\mathcal{S}$ is exponentiable.

(S2) ("Representability Axiom") There exists a map $\pi : E \to U$ in $\mathcal{S}$ which is universal in the following sense: for any map $f : Y \to X$ in $\mathcal{S}$ there exists a diagram

$$
\begin{array}{ccccc}
Y & \longleftarrow & Y' & \longrightarrow & E \\
{\scriptstyle f}\downarrow & & {\scriptstyle f'}\downarrow & & \downarrow{\scriptstyle \pi} \\
X & \underset{p}{\longleftarrow} & X' & \underset{c}{\longrightarrow} & U
\end{array}
\tag{4}
$$

in which p is epi and both squares are pullbacks.

Note that this Representability Axiom states that every map in $\mathcal{S}$ is "locally" a pullback of the universal map $\pi : E \to U$.

From now on, we will refer to a map $f : Y \to X$ in $\mathcal{S}$ as a "small map", or as a "small object over X". Furthermore, an object Y of $\mathcal{C}$ is said to be "small" if the unique map $Y \to 1$ belongs to $\mathcal{S}$.

In the rest of this section, we will make some elementary first observations concerning these axioms for small maps. First notice the following closure properties for exponentiable maps.

1.2 Lemma. *In any Heyting pretopos $\mathcal{C}$, the class of exponentiable maps satisfies the axioms (A1 - 6) for open maps.*

Proof. Write $\mathcal{E}$ for the class of exponentiable maps. First recall that a map $f : Y \to X$ is exponentiable iff the pullback functor $f^* : \mathcal{C}/X \to \mathcal{C}/Y$ has a right adjoint Π_f. From this it is clear that the class $\mathcal{E}$ satisfies axiom (A1). Axioms (A4) and (A5) also clearly hold, the exponential of a sum being constructed as a product (of exponentials). For (A6), it suffices to consider the case $B = 1$ (replace $\mathcal{C}$ by $\mathcal{C}/B$). But for any epimorphism $p : Y \twoheadrightarrow X$, any exponential A^X can be constructed from the exponential A^Y using the universal quantifiers in $\mathcal{C}$, as $A^X = \{f \in A^Y | \forall y_1, y_2 \in Y(p(y_1) = p(y_2) \Rightarrow f(y_1) = f(y_2))\}$. For (A2) assume again that $X = 1$. Then for any exponentiable object Y, any exponential $(A \to X')^{(Y \times X' \to X')}$ in $\mathcal{C}/X'$ can be constructed from the transpose $X' \to X'^Y$ of the projection, as the pullback

Finally, we outline the proof for the descent axiom (A3). We may assume again that $X = 1$. Suppose $X' \twoheadrightarrow 1$ is epi and $Y \times X' \to X'$ is exponentiable in $\mathcal{C}/X'$. Consider an object $A \in \mathcal{C}$, and denote the exponential $(A \times X' \to X')^{(Y \times X' \to X')}$ by $E \to X'$. By axiom (A2), already verified, it follows for the two projections π_1 and $\pi_2 : X' \times X' \rightrightarrows X'$ that the two pullbacks $\pi_i^*(Y \times X' \to X')$ are exponentiable in $\mathcal{C}/X' \times X'$, with exponential $\pi_i^*(E) = \pi_i^*(A \times X' \to X')^{\pi_i^*(Y \times X' \to X')}$ (for $i = 1, 2$). It follows that $E \to X'$ is equipped with canonical descent data. Since in a pretopos every epi is an effective descent map (see Appendix C), it follows that $E \to X'$ is isomorphic to a projection $D \times X' \to X'$, for an object $D \in \mathcal{C}$ uniquely determined up to isomorphism. It is now routine to verify that D is the exponential A^Y.

1.3 Remark. It follows from Lemma 1.2 that the axioms (S1) and (S2) for small maps are equivalent to the single axiom stating that there exists a small map $\pi : E \to U$ which is universal (as in (S2)) as well as exponentiable.

1.4 Remark. Observe that $\pi : E \to U$ is not unique, nor is (given π) the characteristic map c in (4). In fact, for a pullback square

$$
\begin{array}{ccc}
E & \longrightarrow & E' \\
\pi \downarrow & & \downarrow \pi' \\
U & \xrightarrow{f} & U'
\end{array}
$$

with f epi, π is universal iff π' is. However, there is a uniqueness up to "homotopy", just as for universal vector bundles and similar constructions in topology. We refer to the appendix in this chapter (§5) for a precise formulation.

Next, we note the stability of our axioms under slicing. For this, let $\mathcal{S}$ be a class of small maps in $\mathcal{C}$, and let B be an object of $\mathcal{C}$. Define an induced class $\mathcal{S}_B$ in the slice category $\mathcal{C}/B$ in the obvious way: writing $\Sigma_B : \mathcal{C}/B \to \mathcal{C}$ for the forgetful functor, a map f in $\mathcal{C}/B$ belongs to $\mathcal{S}_B$ iff $\Sigma_B(f)$ belongs to $\mathcal{S}$. In the proposition below, $B^* : \mathcal{C} \to \mathcal{C}/B$ denotes the functor $X \mapsto (\pi_2 : X \times B \to B)$, right adjoint to Σ_B.

1.5 Proposition. *Let S be a class of small maps in C. Then S_B is a class of small maps in C/B; moreover, the functor $B^* : C \to C/B$ preserves small maps, as well as the universal small map.*

Proof. The fact that S_B satisfies the axioms for open maps is a matter of elementary verification. Furthermore, S_B clearly satisfies the exponentiability axiom (S1), while (S2) holds for S_B with as universal map $\pi \times B : E \times B \to U \times B$ over B.

To conclude this section, we prove that the notion of "small map" is definable, in the precise sense of the following proposition. As a consequence, one can use the predicate "small" as part of the internal logic of C, as we will freely do in subsequent sections.

1.6 Proposition. *For any arrow f in C over a base B,*

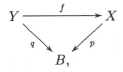

there exists a subobject $S \rightarrowtail B$ such that for any map $\alpha : C \to B$ in C, the pullback $f \times_B C : Y \times_B C \to X \times_B C$ belongs to S iff α factors through S.
This object S will be denoted

$$S = \{b \in B \mid f_b : Y_b \to X_b \text{ is small}\}.$$

Proof. Using exponentiability of the universal small map $\pi : E \to U$, the desired object S can be constructed in terms of the first order logic of C, as

$$S = \{b \in B \mid \forall x \in p^{-1}(b) \ \exists u \in U \ \exists \alpha \in f^{-1}(x)^{\pi^{-1}(u)} : \alpha \text{ is an isomorphism}\}.$$

The verification that this object S has the desired property, stated in the proposition, is straightforward (for example, by using the so-called Kripke-Joyal semantics in C).

§2 Representable structures

In this section we will give some examples of "universal" small structures , obtained from a universal small map $\pi : E \to U$.

As a first case, consider the notion of a universal map between small objects, defined by a representability condition as in axiom (S2). Explicitly, a "universal" map between small objects is a commutative diagram

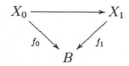

$$(1)$$

in which π_0 and π_1 are small maps, such that for any other such diagram

$$
\begin{array}{ccc}
X_0 & \longrightarrow & X_1 \\
& \searrow f_0 \quad f_1 \swarrow & \\
& B &
\end{array}
$$

with f_0 and f_1 small, there exist an epimorphism $B' \twoheadrightarrow B$ and a map $B' \to V$ which fit into a diagram

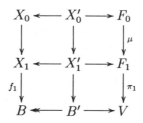

in which all squares are pullbacks.

2.1 Proposition. *There exists a universal map between small objects.*

Proof. Starting from the universal small map $\pi : E \to U$ given by axiom (S2), construct pullbacks $\pi_0^*(E) \to U \times U$ and $\pi_1^*(E) \to U \times U$ of π along the two projections $\pi_0, \pi_1 : U \times U \to U$. Let $V = \pi_1^*(E)^{\pi_0^*(E)} \overset{\alpha}{\to} U \times U$ be the exponential in $\mathcal{C}/U \times U$ (it exists, by axiom (S1)). Define $F_0 = \alpha^* \pi_0^*(E)$ and $F_1 = \alpha^* \pi_1^*(E)$. The evaluation $\varepsilon : \pi_1^*(E)^{\pi_0^*(E)} \times_{(U \times U)} \pi_0^*(E) \to \pi_1^*(E)$ of exponentials in $\mathcal{C}/U \times U$ defines a map $\mu = (1, \varepsilon) : F_0 \to F_1$ over V. One readily verifies that this map is universal, in the sense explained just above the statement of the proposition.

2.2 Corollary. *There exists a universal small map between small objects.*

Proof. Consider a universal map (1) between small objects, given by Proposition 2.1. By definability (Proposition 1.6), one can construct a sub-object $V' = \{v \in V | \mu_v : \pi_0^{-1}(v) \to \pi_1^{-1}(v) \text{ is small}\}$. The pullback of (1) along the inclusion $V' \hookrightarrow V$ is the universal small map between small objects.

As another example, consider for any object A the notion of a "small object labeled by A", i.e. a diagram

$$
\begin{array}{ccc}
Y & \xrightarrow{\lambda} & A \\
\downarrow{\scriptstyle f} & & \\
X & &
\end{array}
\qquad (2)
$$

with $f \in \mathcal{S}$. A universal such labeled small object is such a diagram

$$
\begin{array}{ccc}
E_A & \xrightarrow{\mu} & A \\
\downarrow{\scriptstyle \pi_A} & & \\
U_A & &
\end{array}
\qquad (\pi_A \in \mathcal{S}) \qquad (3)
$$

such that for any other such diagram (2) there exists a commutative diagram

$$
\begin{array}{ccccc}
 & & A & & \\
 & \overset{\lambda}{\nearrow} & \uparrow & \overset{\mu}{\nwarrow} & \\
Y & \longleftarrow & Y' & \longrightarrow & E_A \\
\downarrow{\scriptstyle f} & & \downarrow & & \downarrow{\scriptstyle \pi_A} \\
X & \longleftarrow & X' & \longrightarrow & U_A
\end{array}
\qquad (4)
$$

in which both squares are pullbacks.

2.3 Proposition. *For any object A, there exists a universal small object labeled by A.*

Proof. Starting from the universal small map $\pi : E \to U$ given by the representability axiom, define U_A to be the following exponential in $\mathcal{C}/U$:

$$(U_A \to U) = (A \times U \to U)^{(E \to U)};$$

and define

$$E_A = U_A \times_U E.$$

Let $\pi_A : E_A \to U_A$ be the projection, and let $\mu : E_A \to A$ be the evaluation.

To verify the required universal property, consider any diagram (2) with f small. By universality of $\pi : E \to U$, the map f fits into a double pullback diagram of the form §1 (4). The composition $Y' \to Y \xrightarrow{\lambda} A$ then gives a map $Y' \to A \times U$ over U, i.e. a map $X' \times_U E \to A \times U$ over U. By exponential transposition we obtain a map $X' \to U_A$ over U. This map fits into the following "subdivision" of the diagram §1 (4),

$$
\begin{array}{ccccccc}
Y & \longleftarrow & Y' & \longrightarrow & U_A \times_U E & \longrightarrow & E \\
{\scriptstyle f}\downarrow & & {\scriptstyle f'}\downarrow & & \downarrow & & \downarrow{\scriptstyle \pi} \\
X & \xleftarrow{\ p\ } & X' & \longrightarrow & U_A & \longrightarrow & U
\end{array}
$$

in which the middle square is a pullback, because the rectangle composed out of the two right-hand squares is. With the obvious maps to A, this gives a diagram of the desired form (4). This proves the proposition.

Observe that, given a universal small map $\pi : E \to U$, the explicit construction of a universal labeled object $A \leftarrow E_A \to U_A$ is (covariantly) functorial in A.

For our next example, consider a "directed graph" in $\mathcal{C}$, i.e. a diagram of the form

$$ G_1 \underset{\partial_1}{\overset{\partial_0}{\rightrightarrows}} G_0 . $$

An *action* of this graph on an object $f : Y \to X$ over X consists of a map $e : Y \to G_0$, together with an "action" map $a : \partial_1^*(Y) \to \partial_0^*(Y)$ over X as well as over G_1. (Here $\partial_i^*(Y)$ is the pullback $Y \times_{G_0} G_1 \to G_1$ of e and ∂_i, with obvious map to X.) Briefly, we call such a structure (f, e, a) a *G-object over* X. It is said to be *small* if the map f is small.

If $g : X' \to X$ is any map in $\mathcal{C}$, then from such a small G-object $Y = (f : Y \to X, e, a)$ over X one can construct by pullback a small G-object $g^*(Y)$ over X' in the obvious way. A small G-object $D = (p : D \to W, \varepsilon, \alpha)$ is said to be *universal* if for any other small G-object $Y = (f : Y \to X, e, a)$ there exists an epimorphism $g : X' \to X$ and a map $c : X' \to W$ such that there is an isomorphism $g^*(Y) \cong c^*(D)$ of G-objects over X'.

2.4 Proposition. *For a graph G with small codomain $\partial_1 : G_1 \to G_0$, there exists a universal small G-object.*

Proof. The desired universal object is constructed as follows. Start with the universal small object labeled by G_0,

$$E_{G_0} \xrightarrow{\ \lambda\ } G_0$$
$$\pi_{G_0} \downarrow$$
$$U_{G_0},$$

constructed in Proposition 2.3. From $\lambda : E_{G_0} \to G_0$, one obtains objects $\partial_0^*(E_{G_0})$ and $\partial_1^*(E_{G_0})$ over U_{G_0}. This last object is small over U_{G_0}, since by assumption $\partial_1 : G_1 \to G_0$ is a small map. Thus one can form the exponential

$$V = \partial_0^*(E_{G_0})^{\partial_1^*(E_{G_0})} \to U_{G_0}$$

in $\mathcal{C}/U_{G_0}$. Define $C = V \times_{U_{G_0}} E_{G_0}$, with projection $p : C \to V$. This object C is equipped with an obvious map $\varepsilon : C \to G_0$, and a map $\partial_1^*(C) \xrightarrow{\alpha} \partial_0^*(C)$ over V. Now use the first order logic of the category $\mathcal{C}$ to construct a subobject $W \subseteq V$ consisting of those $v \in V$ for which the map α restricts for the fiber $p^{-1}(v) = C_v$ to a map $\partial_1^*(C_v) \to \partial_0^*(C_v)$ over G_1. Let $D = W \times_V C$ be the restriction of C to this subobject W. Then p and ε, α restrict to maps $p : D \to W$, $\varepsilon : D \to G_0$ and $\alpha : \partial_1^*(D) \to \partial_0^*(D)$ over $W \times G_1$. Thus $D \to W$ has the structure of a G-object over W, small because p is a small map. The verification of the desired universal property of $(D \to W, \varepsilon, \alpha)$ is again routine, and left to the reader.

2.5. Corollary. *There exists a universal G-object with small action map.*

Proof. By definability of smallness (Proposition 1.6), the universal G-object $(D \to W, \varepsilon, \alpha)$ can be restricted to the subobject $W' \subseteq W$ defined by

$$W' = \{w \in W \mid \alpha_w : \partial_1^*(D_w) \to \partial_0^*(D_w) \text{ is small}\},$$

to yield the desired universal G-object $D' = D \times_W W'$ with small action map.

2.6 Corollary. *Let G be an internal category in $\mathcal{C}$, with small codomain map $\partial_1 : G_1 \to G_0$. There exists a universal small ($\mathcal{C}$-internal) presheaf on G.*

Proof. Regard G as a graph, and start with the universal G-object $(p : D \to W, \varepsilon, \alpha)$ constructed in Proposition 2.4. The condition that this action satisfies the equations for a presheaf can be expressed in the first order logic of $\mathcal{C}$, using the composition $G_1 \times_{G_0} G_1 \to G_1$ and the identities $G_0 \to G_1$

of the category G. Thus one can construct a subobject $\tilde{W}$,

$$\tilde{W} = \{w \in W \mid \text{the } G\text{-object } D_w = p^{-1}(w) \text{ is a } G\text{-presheaf}\}.$$

The universal small G-presheaf P is the presheaf over $\tilde{W}$ obtained by pull-back, $P = D \times_W \tilde{W}$.

§3   Power-sets

In this section, $\mathcal{S}$ is a fixed class of small maps in an ambient category $\mathcal{C}$ as before.

Let X be any object in $\mathcal{C}$. For a "parameter" object I, an I-indexed family of subobjects of X is a subobject $S \rightarrowtail I \times X$. It is said to be a family of *small* subobjects if the composition $S \rightarrowtail I \times X \to I$ belongs to the class $\mathcal{S}$. Denote by $P^s(X)(I)$ the set of all such I-indexed families of small subobjects of X. Any map $g : J \to I$ in $\mathcal{C}$ induces by pullback an operation

$$g^\# : P^s(X)(I) \to P^s(X)(J),$$

thus making $P^s(X)$ into a contravariant functor $\mathcal{C}^{op} \to Sets$.

3.1 Theorem. *The functor $P^s(X)$ is representable.*

We shall denote the representing object by $P_s(X)$. Thus $P_s(X)$ has the property that there is a bijection, natural in I, between I-indexed families of small subobjects $S \rightarrowtail I \times X$ and arrows $s : I \to P_s(X)$. We call s the characteristic map for S, and denote it by c_S.

Proof of 3.1. Consider the universal small object labelled by X, as constructed in Proposition 2.3,

$$U_X \xleftarrow{\pi_X} E_X \xrightarrow{\lambda} X.$$

Define for each object I in $\mathcal{C}$ an equivalence relation on the arrows $I \to U_X$, by stating that two such arrows $f, g : I \to U_X$ are equivalent iff the maps $f^*(E_X) \to I \times X$ and $g^*(E_X) \to I \times X$, from the two pullbacks of E_X along f and g respectively, have the same image in $I \times X$. This relation is definable in the first order logic of the ambient category $\mathcal{C}$, hence is representable by a subobject $R_X \subseteq U_X \times U_X$; this means that f and g are equivalent iff $(f,g) : I \to U_X \times U_X$ factors through R_X. (In logical notation, $R_X = \{(u, u') \in$

$U_X \times U_X \mid \forall x \in X(\exists e \in \pi_X^{-1}(u)\lambda(e) = x \Leftrightarrow \exists e' \in \pi_X^{-1}(u')\lambda(e') = x)\}.)$
Define $P_s(X)$ to be the coequalizer of the equivalence relation R_X,

$$R_X \rightrightarrows U_X \to P_s(X).$$

To see that this object $P_s(X)$ indeed represents the functor $P^s(X)$, consider a family of small subobjects $S \rightarrowtail I \times X$. Then by universality of $E_X \to U_X$ there is a double pullback diagram of the form

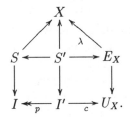

Now for the two projections $\pi_1, \pi_2 : I' \times_I I' \to I'$, the composites $c\pi_1$ and $c\pi_2$ are equivalent (i.e., define the same image in $I' \times_I I' \times X$). Hence, since p is the coequalizer of its kernel pair π_1, π_2, one obtains a unique map c_S making the diagram

$$
\begin{array}{ccc}
I' & \xrightarrow{c} & U_X \\
{\scriptstyle p}\downarrow & & \downarrow \\
I & \dashrightarrow[c_S] & P_s(X)
\end{array}
$$

commute. This defines a map $c_S : I \to P_s(X)$ from $S \rightarrowtail I \times X$. Note that the construction of c_S does not depend on the choice of the cover $I' \twoheadrightarrow I$.

For the converse, any map $f : I \to P_s(X)$ gives by pullback a small map $Z \to I'$ with a map $Z \to X$:

$$
\begin{array}{ccccc}
Z & \longrightarrow & E_X & \longrightarrow & X \\
\downarrow & & \downarrow & & \\
I' & \longrightarrow & U_X & & \\
\downarrow & & \downarrow & & \\
I & \xrightarrow{f} & P_s(X). & &
\end{array}
$$

Let T be the image of Z in $I' \times X$, as in

$$Z \twoheadrightarrow T \rightarrowtail I' \times X.$$

By definition of $P_s(X)$ as the quotient of U_X by R_X, it follows that for the two projections π_1 and $\pi_2 : I' \times_I I' \to I'$, the pullbacks $\pi_1^*(Z)$ and $\pi_2^*(Z)$ have the same image $\pi_1^*(T) \cong \pi_2^*(T)$ in $(I' \times_I I') \times X$. Thus T descends to a subobject $S_f \rightarrowtail I \times X$, fitting into a pullback diagram

$$
\begin{array}{ccc}
T \rightarrowtail I' \times X \longrightarrow I' \\
\downarrow \qquad\quad \downarrow \qquad\quad \downarrow \\
S_f \rightarrowtail I \times X \longrightarrow I.
\end{array}
$$

Since the map $Z \to I'$ is small, it follows by axioms (A6) and (A3) that $S_f \to I$ is also small. This defines a family of small subobjects $S_f \rightarrowtail I \times X$ from the given map $f : I \to P_s(X)$.

We leave it to the reader to verify that these two constructions, of c_S from S and of S_f from f, are mutually inverse.

This completes the proof of Theorem 3.1.

We remark that the representing object $P_s(X)$ of Theorem 3.1 is a covariant functor of X. Indeed, for a subobject $S \rightarrowtail I \times X$ and a map $f : X \to Y$, define $f_!(S)$ to be the image of S under the map $1 \times f : I \times X \to I \times Y$. Then $f_!(S)$ belongs to $P^s(Y)(I)$ whenever S belongs to $P^s(X)(I)$. This operation $f_! : P^s(X)(I) \to P^s(Y)(I)$ is evidently natural in I. By the Yoneda lemma, $f_!$ is given by composition with a uniquely determined map (again denoted)

$$f_! : P_s(X) \to P_s(Y). \tag{1}$$

These maps $f_!$, for all $f : X \to Y$, make $P_s(X)$ into a covariant functor of X.

3.2 Remark. By the stability under slicing of the axioms for small maps (Proposition 1.5), it follows that for any map $X \to B$ in $\mathcal{C}$, there exists an object $P_s(X \to B)$ in $\mathcal{C}/B$ which represents families of small subobjects of $X \to B$ in the slice category $\mathcal{C}/B$. Furthermore, for any arrow $f : A \to B$, the construction of these objects $P_s(X \to B)$ is stable under change of base $f^* : \mathcal{C}/B \to \mathcal{C}/A$.

3.3 Remark. By the Representability Theorem 3.1, there is for each object X a universal family of small subobjects of X, the "membership relation" $\in_X \rightarrowtail P_s(X) \times X$, whose characteristic map is the identity $P_s(X) \to P_s(X)$. The class $\mathcal{S}$ is completely determined by the functor $P_s(X)$ together with all these membership relations $\in_X$, in the sense that a map $f : Y \to X$ belongs to $\mathcal{S}$ iff f is a pullback of $\in_Y \rightarrowtail P_s(Y)$. Indeed, f belongs to $\mathcal{S}$ iff its graph

$(f, 1) : Y \rightarrowtail X \times Y$ belongs to $P^s(Y)(X)$, iff there is a map $f^{-1} : X \rightarrow P_s(Y)$ which fits into a pullback as on the left.

$$
\begin{array}{ccc}
Y & \longrightarrow & \in_Y \\
{\scriptstyle (f,1)}\downarrow & & \downarrow \\
X \times Y & \xrightarrow[f^{-1} \times 1]{} & P_s(Y) \times Y
\end{array}
\qquad\qquad
\begin{array}{ccc}
Y & \longrightarrow & \in_Y \\
{\scriptstyle f}\downarrow & & \downarrow \\
X & \xrightarrow[f^{-1}]{} & P_s(Y)
\end{array}
$$

The square on the left is a pullback iff the one on the right is. Thus, f belongs to $\mathcal{S}$ iff f is a pullback of $\in_Y \rightarrow P_s(Y)$, as claimed.

3.4 Remark. For a small map $X \rightarrow B$, the object $P_s(X \rightarrow B)$ of $\mathcal{C}/B$ acts as a power object for $X \rightarrow B$, for the full subcategory $\mathcal{S}/B$ of $\mathcal{C}/B$ with as objects all small maps into B. Thus this category $\mathcal{S}/B$ is an elementary topos provided that this power object $P_s(X \rightarrow B)$ is itself a small object over B. In Chapter II, §2, we will consider some consequences of the "power-set axiom" which states that $P_s(X \rightarrow B)$ is small whenever $X \rightarrow B$ is.

3.5 Remark. The functor $P_s(-)$ is also *contravariant* along small maps: If $g : Z \rightarrow Y$ belongs to $\mathcal{S}$, then for any element $R \subseteq I \times Y$ of $P^s(Y)(I)$, the pullback $(1 \times g)^{-1}(R) \rightarrowtail I \times Z$ belongs to $P^s(Z)(I)$. Indeed, $I \times Z \rightarrow I \times Y$ is small and hence so is its pullback $(1 \times g)^{-1}(R) \rightarrow R$. Thus the composite $(1 \times g)^{-1}(R) \rightarrow R \rightarrow I$ is small, and $(1 \times g)^{-1}(R)$ belongs to $P^s(Z)(I)$, as claimed.

This operation of pullback is evidently natural in I, hence defines a transformation $g^{-1} : P^s(Y) \rightarrow P^s(Z)$. By the Representability Theorem 3.1, this transformation corresponds to a uniquely determined arrow

$$
g^{-1} : P_s(Y) \rightarrow P_s(Z). \tag{2}
$$

The covariant and contravariant operations (1) and (2) together satisfy the so-called *Beck-Chevalley condition*. This condition states that for a pullback square

$$
\begin{array}{ccc}
X \times_Y Z & \xrightarrow{\bar{f}} & Z \\
{\scriptstyle \bar{g}}\downarrow & & \downarrow{\scriptstyle g} \\
X & \xrightarrow{f} & Y
\end{array}
$$

with g small, the identity

$$
g^{-1} \circ f_! = \bar{f}_! \circ \bar{g}^{-1} : P_s(X) \rightarrow P_s(Z)
$$

holds. (This is an immediate consequence of the fact that in a pretopos, images are stable under pullback.)

3.6 Remark. The covariant and contravariant aspects of the functor P_s can also be described together, using the formalism of bivariant theories of Fulton-MacPherson(1981). Indeed, for a map $f : X \to Y$, let $T(f)$ be the collection of all subobjects $S \subseteq X$ with the property that the composition $S \subseteq X \to Y$ is small. If $f : X \to Y$ and $g : Y \to Z$ are two maps and $S \in T(f)$ while $R \in T(g)$, then there is a "product" $S \cap f^{-1}(R) \in T(g \circ f)$. Furthermore, the image along f defines an operation

$$f_* : T(g \circ f) \to T(g),$$

while for a pullback square as in 3.5 above, there is an obvious pullback operation

$$f^* : T(g) \to T(\bar{g}).$$

With the pullbacks as "independent squares" and all maps "confined", these operations satisfy all the axioms for a (set-valued) bivariant theory as defined in *op. cit.* The associated covariant functor assigns to each object X the collection $Sub_s(X)$ of small objects of X, while the associated contravariant functor assigns to X the collection of all small monomorphisms into X. (In many examples, the class $\mathcal{S}$ of small maps contains all monomorphisms, and this is simply the collection $Sub(X)$ of all subobjects of X.)

To conclude this section, we observe that the maps $f_!$, constructed in (1) on the basis of Theorem 3.1 enable us to give an equivalent formulation of the Collection Axiom (A7). This formulation also illustrates the relation of this axiom to the axiom of Zermelo-Fraenkel set theory bearing the same name.

3.7 Proposition. *The collection axiom (A7) is equivalent (relative to the other axioms) to the condition that for any epimorphism $f : X \to Y$, the induced map $f_! : P_s(X) \to P_s(Y)$ is again epi.*

Proof. ($\Rightarrow$) Consider an epi $f : X \to Y$, and an arbitrary map $c : I \to P_s(Y)$, characteristic for $S \rightarrowtail I \times Y$ with $S \to I$ small. Form the pullback

$$
\begin{array}{ccc}
T & \rightarrowtail & I \times X \\
\downarrow & & \downarrow \\
S & \rightarrowtail & I \times Y.
\end{array}
$$

By the collection axiom, there exists a quasi-pullback square

$$Z \longrightarrow T \twoheadrightarrow S$$
$$\downarrow \qquad\qquad \downarrow$$
$$J \longrightarrow I$$

with $Z \to J$ small. Thus if we write R for the image of Z in $J \times X$, the map $R \to J$ is again small, i.e. $R \in P^s(X)(J)$. Furthermore, since the preceding diagram is a quasi-pullback, the left-hand map in the diagram

$$Z \twoheadrightarrow R \rightarrowtail J \times X$$
$$\downarrow \qquad\qquad\qquad \downarrow {\scriptstyle 1 \times f}$$
$$J \times_I S \rightarrowtail \longrightarrow J \times Y$$

is epi. This shows that $f_!(R) = J \times_I S$. Or in other words, the square

$$J \xrightarrow{\ c_R\ } P_s(X)$$
$$\downarrow \qquad\qquad \downarrow {\scriptstyle f_!}$$
$$I \xrightarrow{\ c\ } P_s(Y)$$

commutes. Since the map c that we started with was arbitrary, this shows that $f_!$ is epi.

($\Leftarrow$) Consider any small map $f : X \to I$ and an epi $p : Y \to X$, as in the formulation of the collection axiom (A7). Then $(f, 1) : X \to I \times X$ defines an element of $P^s(X)(I)$, or an arrow $c : I \to P_s(X)$. Since $p_! : P_s(Y) \to P_s(X)$ is assumed epi, the map $J \to I$ in the pullback diagram

$$J \xrightarrow{\ d\ } P_s(Y)$$
$$\downarrow \qquad\qquad \downarrow$$
$$I \xrightarrow[\ c\ ]{} P_s(X)$$

is again epi. The map d in this diagram is characteristic for a subobject $Z \rightarrowtail J \times Y$. Thus $Z \to J$ is small, and the composite $Z \to J \times Y \to J \times X$ has as its image the pullback of $(f, 1) : X \to I \times X$ along $J \times X \to I \times X$. This pullback is $J \times_I X$, so that

$$Z \longrightarrow Y \longrightarrow X$$
$$\downarrow \qquad\qquad\qquad \downarrow$$
$$J \longrightarrow I$$

is a quasi-pullback, as desired.

This completes the proof.

§4   Complete sup-lattices

Any class of small maps gives rise to a notion of completeness. Consider a category $\mathcal{C}$ equipped with a class of small maps $\mathcal{S}$, as before. Let L be a poset in $\mathcal{C}$. This implies in particular that for each object A of $\mathcal{C}$ the set $\mathcal{C}(A, L)$ of all arrows $A \to L$ is partially ordered. For maps $g : B \to A$ and $\lambda : B \to L$, an arrow $\mu : A \to L$ is said to be *the supremum of λ along g* if for any arrows $t : C \to A$ and $\nu : C \to L$, with associated pullback square

$$
\begin{array}{ccc}
C \times_A B & \xrightarrow{\ \pi_2\ } & B \\
{\scriptstyle \pi_1} \downarrow & & \downarrow {\scriptstyle g} \\
C & \xrightarrow{\ \ t\ \ } & A,
\end{array}
$$

one has

$$\mu \circ t \leq \nu \quad \text{iff} \quad \lambda \circ \pi_2 \leq \nu \circ \pi_1 \quad (\text{in } \mathcal{C}(C \times_A B, L)).$$

Given λ and g, an arrow μ with this property is necessarily unique, and denoted

$$\mu = \bigvee\nolimits_g \lambda : A \to L.$$

(When we use the internal first order logic of $\mathcal{C}$, we also write $\mu(a) = \bigvee_{g(b)=a} \lambda(b)$ for this supremum.) The poset L is said to be $(\mathcal{S}\text{-})complete$ if suprema in L exist along any map in the class $\mathcal{S}$.

Note that by axiom (A4), any such $\mathcal{S}$-complete poset L is a semi-lattice: the smallest element $\perp : 1 \to L$ is the supremum along $0 \to 1$ of the unique map $0 \to L$, while the binary supremum $\vee : L \times L \to L$ is constructed as the supremum of the map $\binom{\pi_1}{\pi_2} : (L \times L) + (L \times L) \to L$ along the map $\binom{id}{id} : (L \times L) + (L \times L) \to (L \times L)$ (which, being a pullback of the map $1 + 1 \to 1$, is small).

4.1 Example. For any object X in $\mathcal{C}$, the poset $P_s(X)$ (with order induced by inclusion of subobjects) is $\mathcal{S}$-complete. Indeed, for a map $\lambda : B \to P_s(X)$ and a small map $g : B \to A$, one constructs the supremum $\bigvee_g \lambda$ of λ along g as follows: λ is the characteristic map of a subobject $R \rightarrowtail B \times X$ with the property that $R \to B$ is small. Let S be the image of $g \times 1$, as in

$$
\begin{array}{ccccc}
R & \rightarrowtail & B \times X & \longrightarrow & B \\
\downarrow & & \downarrow & & \downarrow {\scriptstyle g} \\
S & \rightarrowtail & A \times X & \longrightarrow & A.
\end{array}
$$

Then $S \to A$ is small since $R \to B$ and $B \to A$ are both small. Thus S has a characteristic map $A \to P_s(X)$; this map is the required supremum.

4.2 Proposition. *$P_s(X)$ is the free S-complete sup-lattice generated by X.*

For the proof, write

$$\{\cdot\} : X \to P_s(X) \quad \text{("singleton")} \tag{1}$$

for the characteristic map of the diagonal $X \rightarrowtail X \times X$. Then any map $g : X \to L$ into an S-complete sup-lattice L can be extended uniquely to a map $\bar{g}$ which preserves suprema along small maps:

$$
\begin{array}{ccc}
X & \xrightarrow{\ g\ } & L \\
{\scriptstyle \{\cdot\}} \downarrow & \nearrow {\scriptstyle \bar{g}} & \\
P_s(X). & &
\end{array}
$$

Indeed, for $r : I \to P_s(X)$, characteristic for $R \subseteq I \times X$ with $R \to I$ small, let

$$(\bar{g} \circ r) := \bigvee\nolimits_{\pi_1} (g\pi_2 : R \to L),$$

where π_1 and π_2 are the projections $R \to I$ and $R \to X$. This formula, for all maps r, defines the extension $\bar{g}$. We leave further details to the reader. Observe that $P_s(X)$ is also free in a stronger sense which includes arbitrary parameters (as, for example, in our description of the natural numbers object, cf. Appendix B).

4.3 Remark. By freeness of $P_s(X)$, there is a canonical union operation $\cup : P_s P_s(X) \to P_s(X)$, which extends the identity on $P_s(X)$ in this way:

$$
\begin{array}{ccc}
P_s(X) & \xrightarrow{\ id\ } & P_s(X) \\
{\scriptstyle \{\cdot\}} \downarrow & \nearrow {\scriptstyle \cup} & \\
P_s P_s(X). & &
\end{array}
$$

The maps $\{\cdot\} : X \to P_s(X)$ and $\cup : P_s P_s(X) \to P_s(X)$ give the "powerset" functor $P_s(-)$ the structure of a monad. For some applications of this monad structure, see Appendix A.

4.4 Remark. Although we will not use this fact in the present book, it may be of interest to point out that the class $\mathcal{S}$ can be recovered from its $\mathcal{S}$-complete semi-lattices, in the sense that a map $g : B \to A$ belongs to $\mathcal{S}$ iff suprema along g exist in any $\mathcal{S}$-complete semi-lattice.

§5   Appendix: Uniqueness of universal small maps

As observed in §1, the universal small map $\pi : E \to U$ is not unique, and there is some choice involved in the construction of characteristic maps c (cf. diagram (4) of §1). However, there is a uniqueness "up to homotopy", as for classifying spaces for vector bundles in topology. This homotopy uniqueness can be expressed by means of groupoids in $\mathcal{C}$.

Let $Gpd(\mathcal{C})$ denote the 2-category of internal groupoids, functors and natural transformations in $\mathcal{C}$. For an object G of $Gpd(\mathcal{C})$, we denote its "space" of objects by G_0, its space of arrows by G_1, and its domain and codomain by ∂_0 and ∂_1, respectively,

$$G_1 \underset{\partial_1}{\overset{\partial_0}{\rightrightarrows}} G_0.$$

Every object X of $\mathcal{C}$ defines a "discrete groupoid"

$$X_{\mathrm{dis}} = (X \underset{id}{\overset{id}{\rightrightarrows}} X)$$

with X as space of objects, and only identity arrows. This defines an embedding

$$\mathcal{C} \to Gpd(\mathcal{C}).$$

For a map $f : Y \to X$ in $\mathcal{C}$, its kernel pair $Y \times_X Y \rightrightarrows Y$ is an equivalence relation on Y, and can be viewed as a groupoid in $\mathcal{C}$, with Y as space of objects and $Y \times_X Y$ as space of arrows. We call this groupoid the kernel of f and denote it $Ker(f)$.

If the map f is exponentiable, there is another groupoid associated to f, with X as space of objects. Using the first order logic of $\mathcal{C}$, the arrows $x \to x'$ in this groupoid can be described as the isomorphisms between the fibers $f^{-1}(x) \overset{\sim}{\to} f^{-1}(x')$. We will denote this groupoid by $(Iso(f) \rightrightarrows X)$, or simply by $Iso(f)$. The space of arrows of this groupoid can also be described categorically, in the standard way: consider the pullbacks $\pi_1^*(Y)$ and $\pi_2^*(Y)$ of $f : Y \to X$ along the projections π_1 and $\pi_2 : X \times X \to X$, and form the exponential $\pi_2^*(Y)^{\pi_1^*(Y)} \to X \times X$ in $\mathcal{C}/(X \times X)$. Then the space of arrows of the groupoid is the subspace $Iso(f) \subseteq \pi_2^*(Y)^{\pi_1^*(Y)}$ consisting of isomorphisms.

Recall that a homomorphism (internal functor) $\varphi : G \to H$ between internal groupoids in $\mathcal{C}$ is said to be *essentially surjective* if the map $\partial_1 \pi_1$ in the diagram below is epi:

$$
\begin{array}{ccccc}
H_1 \times_{H_0} G_0 & \xrightarrow{\pi_1} & H_1 & \xrightarrow{\partial_1} & H_0 \\
\downarrow & & \downarrow{\scriptstyle \partial_0} & & \\
G_0 & \longrightarrow & H_0. & &
\end{array}
$$

Furthermore, φ is said to be *fully faithful* if the square

$$
\begin{array}{ccc}
G_1 & \longrightarrow & H_1 \\
\downarrow & & \downarrow \\
G_0 \times G_0 & \longrightarrow & H_0 \times H_0
\end{array}
$$

is a pullback. A *weak equivalence* is a homomorphism $\varphi : G \to H$ which is both essentially surjective and fully faithful. For example, if $f : Y \to X$ is an epimorphism then there is an obvious weak equivalence, denoted

$$
\bar{f} : Ker(f) \xrightarrow{\sim} X_{\text{dis}}.
$$

Let $\mathcal{W}$ be the collection of all weak equivalences. The homotopy category $HoGpd(\mathcal{C})$ is the category obtained from $Gpd(\mathcal{C})$ by inverting all weak equivalences:

$$
HoGpd(\mathcal{C}) = Gpd(\mathcal{C})[\mathcal{W}^{-1}].
$$

This category can be constructed, for example, by first passing from the 2-category $Gpd(\mathcal{C})$ to the category $\pi_0 Gpd(\mathcal{C})$ of groupoids and isomorphism classes of internal functors. Then the image $\pi_0(\mathcal{W})$ of $\mathcal{W}$ admits a calculus of (right) fractions in $\pi_0 Gpd(\mathcal{C})$ (in the sense of Gabriel-Zisman(1967)), and $HoGpd(\mathcal{C})$ is the category of fractions $\pi_0(Gpd(\mathcal{C}))[\pi_0(\mathcal{W})^{-1}]$. For two groupoids G and H in $\mathcal{C}$, we denote by

$$
[G, H]
$$

the collection of arrows $G \to H$ in this homotopy category. Thus any element u of $[G, H]$ is represented by a diagram

$$
G \xleftarrow{w} K \xrightarrow{\varphi} H \ , \ u = \varphi \circ w^{-1}
$$

where the first arrow $K \to G$ is a weak equivalence.

Now consider "the" universal small map $\pi : E \to U$ given by the representability axiom (S2) in §1. Let f be a small map, with characteristic map c defined on a cover p, as in the double pullback diagram

$$
\begin{array}{ccc}
Y & \longleftarrow Y' \longrightarrow & E \\
f\downarrow & \quad\downarrow f' & \quad\downarrow \pi \\
X & \underset{p}{\longleftarrow} X' \underset{c}{\longrightarrow} & U.
\end{array}
$$

Then for each pair of points $(z_1, z_2) \in X' \times_X X'$, this diagram gives a specific isomorphism of fibers

$$
\alpha_{z_1, z_2} : E_{c(z_1)} = \pi^{-1}(c(z_1)) \cong Y'_{z_1} \cong Y_{p(z_1)} = Y_{p(z_2)} \cong Y'_{z_2} \cong E_{c(z_2)}.
$$

Thus one obtains a homomorphism of groupoids

$$
\alpha : Ker(p) \to Iso(\pi).
$$

Together with the weak equivalence

$$
\bar{p} : Ker(p) \to X_{\text{dis}},
$$

one thus obtains a map $\alpha \circ \bar{p}^{-1} : X_{\text{dis}} \to Iso(\pi)$ in $HoGpd(\mathcal{C})$. This construction gives an operation

$$
\{\text{isomorphism classes of small maps over } X\} \to [X_{\text{dis}}, Iso(\pi)]. \qquad (1)
$$

5.1 Proposition. *This operation (1) is a natural bijection.*

Proof. One readily checks that the operation is natural in X. To show it is bijective, we explicitly describe an inverse operation. Consider a map $u : X_{dis} \to Iso(\pi)$ in the homotopy category. Then u can be represented as $u = \varphi \circ w^{-1}$,

$$
X_{\text{dis}} \xleftarrow{w} G \xrightarrow{\varphi} Iso(\pi),
$$

where w is a weak equivalence. Thus $w : G_0 \to X$ is an epimorphism while $G_1 \cong G_0 \times_X G_0$; or in other words, $G \cong Ker(w : G_0 \twoheadrightarrow X)$. Consider the map $\varphi : G_0 \to U$ on objects, and construct the pullback

$$
\begin{array}{ccc}
P & \longrightarrow & E \\
\downarrow & & \quad\downarrow \pi \\
G_0 & \underset{\varphi}{\longrightarrow} & U.
\end{array}
$$

The map φ on arrows gives for each arrow $g : z_1 \to z_2$ in G_1 an isomorphism $\varphi(g) : E_{\varphi(z_1)} \xrightarrow{\sim} E_{\varphi(z_2)}$, thus an isomorphism $\varphi(g) : P_{z_1} \to P_{z_2}$. Since $G \cong Ker(G_0 \to X)$, these isomorphisms together provide the map $P \to G_0$ with *descent data* for the epimorphism $w : G_0 \twoheadrightarrow X$. Since in a pretopos every epimorphism is an effective descent map (see Appendix B), it follows that there is a map $f : Y \to X$ which fits into a pullback square

$$
\begin{array}{ccc}
Y & \longleftarrow & P \\
{\scriptstyle f}\downarrow & & \downarrow \\
X & \longleftarrow & G_0
\end{array}
\tag{2}
$$

and is compatible with the descent data on P. It follows by the axioms (A2) and (A3) for open maps in §1 that f is small, since π is.

It is straightforward to verify that this construction, of $f : Y \to X$ from the map u in the homotopy category, does not depend on the chosen representation $u = \varphi \circ w^{-1}$ (any other representation would have produced a small map isomorphic to f), and that this construction provides a two-sided inverse to the operation (1).

5.2 Corollary. *The universal small map* $\pi : E \to U$ *is unique up to weak equivalence of the associated groupoid* $Iso(\pi) \rightrightarrows U$.

In other words, if $\pi' : E' \to U'$ is any other universal small map, there exist a groupoid G and a pair of weak equivalences $Iso(\pi) \xleftarrow{\sim} G \xrightarrow{\sim} Iso(\pi')$.

Chapter II

Zermelo-Fraenkel Algebras

Throughout this chapter, $\mathcal{S}$ is a fixed class of small maps in a Heyting pretopos $\mathcal{C}$, as defined in Chapter I, §1.

§1 Free Zermelo-Fraenkel algebras

We begin by introducing the basic notion in this chapter.

1.1 Definition. A *Zermelo-Fraenkel (ZF) algebra* in $\mathcal{C}$ is an $\mathcal{S}$-complete sup-lattice L in $\mathcal{C}$ equipped with a map $s : L \to L$, called the *successor* operation . A *homomorphism* of such algebras $(L, s) \to (M, t)$ is a map $f : L \to M$ which preserves suprema along small maps, and commutes with successors.

In any such ZF-algebra (L, s), it is possible to define a *"membership relation"* $\varepsilon \subseteq L \times L$ by setting, for (generalized) elements x, y of L:

$$x \varepsilon y \quad \text{iff} \quad s(x) \leq y. \tag{1}$$

Any homomorphism preserves this membership relation.

Observe that if (L, s) is a ZF-algebra in $\mathcal{C}$, then for any object B the pullback $B^*(L, s) = (L \times B \to B)$ is a ZF-algebra in the slice category $\mathcal{C}/B$, with respect to the induced class $\mathcal{S}_B$ of small maps; cf. Proposition I.1.5.

We shall be particularly interested in free ZF-algebras . Since $\mathcal{C}$ need not be cartesian closed, the freeness condition should be formulated with arbitrary parameters (as we did for a natural numbers object in a pretopos, cf. Appendix B). Thus, for an object A in $\mathcal{C}$, a free ZF-algebra on A is such an algebra $V(A)$ equipped with a map $\eta : A \to V(A)$, with the property

29

that for any object B in $\mathcal{C}$ and any ZF-algebra (L, s) in $\mathcal{C}/B$, any map $\varphi : B^*(A) \to L$ in $\mathcal{C}/B$ can be uniquely extended to a homomorphism of ZF-algebras $\bar{\varphi} : B^*(V(A)) \to L$ in $\mathcal{C}/B$:

$$
\begin{array}{ccc}
A \times B & \xrightarrow{\;\varphi\;} & L \\
{\scriptstyle \eta \times B} \big\downarrow & \nearrow {\scriptstyle \bar{\varphi}} & \\
V(A) \times B. & &
\end{array}
$$

When $A = 0$ we write V for $V(A)$. The algebra $V(A)$ is called the cumulative hierarchy on A.

In this section, we will derive some properties of these free algebras $V(A)$. In the next chapter, we will show that these free algebras exist in the category $\mathcal{C}$.

We begin by considering the free ZF-algebra V, and show that, up to isomorphism, this is always an algebra of "small sets":

1.2 Theorem. *The map* $r : P_s(V) \to V$ *defined by the formula*

$$
r(E) = \bigvee\nolimits_{x \in E} s(x)
$$

is an isomorphism of $\mathcal{S}$-complete sup-lattices.

Proof. In this proof, as elsewhere, we shall exploit the first order logic of $\mathcal{C}$ and use set-theoretic notation to describe arrows in $\mathcal{C}$, in the usual way. We will also distinguish between the categorical membership $\in$ for generalized elements (defined in any category), and the formal membership relation ε in a ZF-algebra, defined in (1) above.

For the proof, we first use the singleton map $\{\cdot\} : V \to P_s(V)$ (see §I. 4, (1)) to define a successor operation

$$
s' : P_s(V) \to P_s(V)
$$

by

$$
s'(E) = \{\bigvee\nolimits_{x \in E} s(x)\}.
$$

Since $P_s(V)$ is a complete sup-lattice (cf. §I.4), this defines a ZF-algebra $(P_s(V), s')$. Since V is the free ZF-algebra, there is thus a unique homomorphism of ZF-algebras

$$
i : V \to P_s(V).
$$

On the other hand, there is the map $r : P_s(V) \to V$ as defined in the statement of the theorem. This map r clearly preserves small suprema, and it preserves the successor by the very definition of s'. In other words, r is a homomorphism of ZF-algebras. But then $r \circ i = id : V \to V$, by freeness of V. Or more explicitly, for any $v \in V$ the identity

$$v = \bigvee_{x \in i(v)} s(x) \tag{2}$$

holds. But then also

$$s'i(v) = \{\bigvee_{x \in i(v)} s(x)\} = \{v\}. \tag{3}$$

Hence for any $E \in P_s(V)$,

$$
\begin{aligned}
ir(E) &= i(\bigvee_{x \in E} s(x)) \\
&= \bigvee_{x \in E} is(x) \\
&= \bigvee_{x \in E} s'i(x) \\
&= \bigvee_{x \in E} \{x\} \qquad \text{(by (3))} \\
&= E.
\end{aligned}
$$

This shows that i and r are mutually inverse isomorphisms, and completes the proof.

1.3 Remark. In Appendix A we present an abstract version of this proof, which shows that Theorem 1.2 is in fact a special instance of a general property of algebras for a monad.

We single out some immediate consequences of Theorem 1.2 and its proof.

1.4. Corollary. *The free algebra (V, s) has the following properties:*

(i) $\forall x, y \in V : sx \le sy \Rightarrow x = y$.

(ii) $\forall y \in V : \{x | x \varepsilon y\}$ *is small (in other words, $\pi_2 : \{(x,y) \in V \times V | x \varepsilon y\} \to V$ belongs to S) and $y = \bigvee_{x \varepsilon y} s(x)$.*

(iii) *(Irreducibility of successors) For any $y \in V$ and any $E \in P_s(V)$: if $sy \le \bigvee_{x \in E} x$ then $\exists x \in E(sy \le x)$.*

Proof. (i) follows from (3) above and the identity $s'i = is$; indeed, if $sx \leq sy$ then $isx \leq isy$, or $s'ix \leq s'iy$, whence by (3) $\{x\} \subseteq \{y\}$, which implies $x = y$.

For (ii), first observe that for any $x, y \in V$,

$$x \in i(y) \quad \text{iff} \quad \{x\} \leq i(y)$$

$$\text{iff} \quad s'i(x) \leq i(y) \qquad \text{(by (3))}$$

$$\text{iff} \quad is(x) \leq i(y)$$

$$\text{iff} \quad s(x) \leq y$$

$$\text{iff} \quad x\varepsilon y.$$

Hence, since the identity $E = \bigcup_{x \in E}\{x\} = \{x | x \in E\}$ holds for any $E \in P_s(V)$, also

$$i(y) = \{x | x\varepsilon y\}, \tag{4}$$

and in particular, the right-hand side of this equation is small. Now apply the isomorphism r to (4), to get

$$y = ri(y) \;\; = \;\; \bigvee_{x\varepsilon y} r(\{x\})$$

$$= \;\; \bigvee_{x\varepsilon y} s(x).$$

This proves (ii).

Finally, for (iii), assume $sy \leq \bigvee_{x \in E} x$. Then by applying the isomorphism i, and using (3) and (4), we find that in $P_s(V)$

$$\{y\} = s'i(y) = is(y) \;\; \leq \;\; i(\bigvee_{x \in E} x)$$

$$= \;\; \bigvee_{x \in E} i(x)$$

$$= \;\; \bigvee_{x \in E} \{z | z\varepsilon x\}$$

$$= \;\; \{z | \exists x\varepsilon E(z\varepsilon x)\}.$$

Thus $\exists x \in E(y\varepsilon x)$, as required.

The analog of Theorem 1.2 for the cumulative hierarchy on an object $A \in C$ reads:

1.5 Theorem. *The map* $r : P_s(A) \times P_s(V(A)) \to V(A)$ *defined by*

$$r(U, E) = \bigvee_{a \in U} \eta(a) \vee \bigvee_{x \in E} s(x)$$

is an isomorphism of S-*complete sup-lattices.*

Proof. Notice first that if we define the map

$$\eta' : A \to P_s(A) \times P_s(V(A))$$

by

$$\eta'(a) = (\{a\}, \emptyset),$$

then η' corresponds to η under the intended isomorphism r, in the sense that $\eta = r \circ \eta'$. In addition to this map η', equip the sup-lattice $P_s(A) \times P_s(V(A))$ with a successor

$$s' : P_s(A) \times P_s(V(A)) \to P_s(A) \times P_s(V(A))$$

defined by

$$
\begin{aligned}
s'(U, E) &= (\emptyset, \{\textstyle\bigvee_{a \in U} \eta(a) \vee \bigvee_{x \in E} s(x)\}) \\
&= (\emptyset, \{r(U, E)\}).
\end{aligned}
$$

Then by freeness of $V(A)$, there is a unique homomorphism of ZF-algebras

$$i : V(A) \to P_s(A) \times P_s(V(A))$$

with the property that $i \circ \eta = \eta'$.

In the reverse direction, consider the map $r : P_s(A) \times P_s(V(A)) \to V(A)$ as defined in the statement of the theorem. This map r clearly preserves sups, and satisfies the equation $s \circ r = r \circ s'$; thus r is a homomorphism of ZF-algebras. Furthermore, $r \circ \eta' = \eta$, as observed above. Thus the composite $r \circ i$ is a homomorphism $V(A) \to V(A)$ which respects $\eta : A \to V(A)$, hence must be the identity. In other words, for any $v \in V(A)$, for which we can write $i(v) = (i_1(v), i_2(v))$, the identity

$$v = \bigvee_{a \in i_1(v)} \eta(a) \vee \bigvee_{x \in i_2(v)} s(x) \tag{5}$$

holds. Furthermore, by definition of s',

$$s'(v) = (\emptyset, \{ri(v)\})$$

$$= (\emptyset, \{v\}).$$

(6)

Thus for any $E \in P_s(V(A))$,

$$ir(\emptyset, E) = i(\bigvee_{x \in E} s(x))$$

$$= \bigvee_{x \in E} is(x)$$

$$= \bigvee_{x \in E} s'i(x)$$

$$= \bigvee_{x \in E} (\emptyset, \{x\}) \quad \text{(by (6))}$$

$$= (\emptyset, E).$$

And similarly, for any $U \in P_s(A)$,

$$ir(U, \emptyset) = i(\bigvee_{a \in U} \eta(a))$$

$$= \bigvee_{a \in U} i\eta(a)$$

$$= \bigvee_{a \in U} \eta'(a)$$

$$= \bigvee_{a \in U} (\{a\}, \emptyset)$$

$$= (U, \emptyset).$$

But then, for an arbitrary $(U, E) \in P_s(A) \times P_s(V(A))$,

$$ir(U, E) = ir((U, \emptyset) \vee (\emptyset, E))$$

$$= ir(U, \emptyset) \vee ir(\emptyset, E)$$

$$= (U, \emptyset) \vee (\emptyset, E)$$

$$= (U, E).$$

This shows that ir is the identity map. Thus i and r are mutually inverse isomorphisms, and the theorem is proved.

1.6 Remark. This theorem expresses that $V(A)$ is an algebra of pairs of small sets (U, E), where U is a set of "atoms" (elements of A) and E is a set of such pairs of small sets. In set theory, one usually defines a universe $V'(A)$, constructed from a collection of atoms of A, to consist of atoms, or sets of elements of $V'(A)$. In other words, $V'(A)$ satisfies the "recursion equation"

$$V'(A) \cong A + P_s(V'(A)).$$

One obtains a solution to this equation from the free algebra $V(A)$, by defining $V'(A) := A + V(A)$. Indeed, by the theorem, one then has

$$A + P_s(V'(A)) \;= A + P_s(A + V(A))$$

$$\cong A + (P_s(A) \times P_s(V(A)))$$

$$\cong A + V(A)$$

$$= V'(A).$$

Observe that for $A = 0$ we have $V'(A) = V(A) = V$. We will show in §5 below that under suitable hypotheses, $V'(A)$ is a model of intuitionistic Zermelo-Fraenkel set theory.

The analog of Corollary 1.4 for $V(A)$ is the following:

1.7 Corollary. *The free algebra $V(A)$ on A has the following properties:*
(i) *For any $x, y \in V(A)$, $s(x) \leq s(y)$ implies $x = y$.*

(ii) *For any $y \in V(A)$, the subobjects $\{a | \eta(a) \leq y\} \subseteq A$ and $\{x | x \varepsilon y\} \subseteq V(A)$ are small, and*

$$y = [\textstyle\bigvee_{\eta(a) \leq y} \eta(a) \vee \bigvee_{x \varepsilon y} s(x)].$$

(iii) *(Irreducibility of generators and successors) For any $a \in A$, any $y \in V(A)$ and any $E \in P_s(V(A))$:*

$$if\ \eta(a) \leq \textstyle\bigvee_{x \in E} x\ then\ \exists x \in E(\eta(a) \leq x);$$
$$if\ s(y) \leq \textstyle\bigvee_{x \in E} x\ then\ \exists x \in E(s(y) \leq x).$$

Proof. (i) For (U, E) and (V, F) in $P_s(A) \times P_s(V(A))$, one has

$$s'(U, E) \leq s'(V, F) \quad \text{iff} \quad (\emptyset, \{r(U, E)\}) \leq (\emptyset, \{r(V, F)\})$$

$$\text{iff} \quad \{r(U, E)\} \leq \{r(V, F)\}$$

$$\text{iff} \quad r(U, E) = r(V, F)$$

$$\text{iff} \quad (U, E) = (V, F),$$

the last since r is an isomorphism. In particular, for any $x, y \in V(A)$,

$$s(x) \leq s(y) \quad \text{iff} \quad is(x) \leq is(y)$$

$$\text{iff} \quad s'i(x) \leq s'i(y)$$

$$\text{iff} \quad i(x) = i(y)$$

$$\text{iff} \quad x = y.$$

This proves part (i).

For (ii), write $i(y) = (i_1(y), i_2(y))$, as before. Then for any $a \in A$,

$$a \in i_1(y) \quad \text{iff} \quad \{a\} \leq i_1(y)$$

$$\text{iff} \quad (\{a\}, \emptyset) \leq i(y)$$

$$\text{iff} \quad r(\{a\}, \emptyset) \leq y$$

$$\text{iff} \quad \eta(a) \leq y.$$

Or in other words,

$$i_1(y) = \{a \in A \,|\, \eta(a) \leq y\}. \tag{7}$$

In particular, the right-hand side of (7) is small.

Similarly, for any $x \in V(A)$,

$$x \in i_2(y) \quad \text{iff} \quad \{x\} \le i_2(y)$$

$$\text{iff} \quad (\emptyset, \{x\}) \le i(y)$$

$$\text{iff} \quad r(\emptyset, \{x\}) \le y$$

$$\text{iff} \quad s(x) \le y \qquad (\text{i.e., } x\varepsilon y).$$

Thus

$$i_2(y) = \{x \in V(A) \mid s(x) \le y\}, \tag{8}$$

and the right-hand side is again small. Now we can write

$$y = ri(y) \quad = r(i_1(y), i_2(y))$$

$$= \bigvee_{a \in i_1(y)} \eta(a) \vee \bigvee_{x \in i_2(y)} s(x) \quad (\text{def. of } r)$$

$$= \bigvee_{\eta(a) \le y} \eta(a) \vee \bigvee_{x\varepsilon y} s(x) \qquad (\text{by } (7), (8)),$$

and (ii) is proved.

Finally, we prove (iii). For y, a, E as in the statement of part (iii) of the corollary,

$$s(y) \le \bigvee_{x \in E} x \quad \text{iff} \quad is(y) \le \bigvee_{x \in E} i(x)$$

$$\text{iff} \quad s'i(y) \le \bigvee_{x \in E} i(x)$$

$$\text{iff} \quad (\emptyset, \{y\}) \le \bigvee_{x \in E} i(x) \qquad (\text{by } (6))$$

$$\text{iff} \quad \{y\} \le \bigvee_{x \in E} i_2(x)$$

$$\text{iff} \quad \{y\} \le \{z \mid \exists x \in E \ (s(z) \le x)\} \quad (\text{by } (8))$$

$$\text{iff} \quad \exists x \in E \ s(y) \le x.$$

Similarly,

$$\eta(a) \leq \bigvee_{x \in E} x \quad \text{iff} \quad \eta'(a) = i\eta(a) \leq \bigvee_{x \in E} i(x)$$

$$\text{iff} \quad (\{a\}, \emptyset) \leq \bigvee_{x \in E} i(x) \qquad (\text{def. of } \eta')$$

$$\text{iff} \quad \{a\} \leq \bigvee_{x \in E} i_1(x)$$

$$\text{iff} \quad \exists x \in E \ \eta(a) \leq x \qquad (\text{by } (7)).$$

This proves part (iii).

§2   Ordinal numbers

In this section we consider the free ZF-algebra with a *monotone* successor, and denote it by (O, t). Thus O is an $\mathcal{S}$-complete sup-lattice, the successor $t : O \to O$ satisfies

$$x \leq y \ \Rightarrow \ tx \leq ty,$$

and (O, t) is the initial ZF-algebra with these properties. (Moreover, since freeness is to be interpreted with arbitrary parameters, (O, t) remains initial when pulled back to any slice category of $\mathcal{C}$.) We shall call (O, t) the *algebra of ordinal numbers* .

As a first observation, we note that the successor t is inflationary:

2.1 Proposition. *In the algebra (O, t) of ordinal numbers, the inequality $x \leq t(x)$ holds.*

Proof. Let $A = \{x \in O | x \leq tx\}$. We show that A is closed under small suprema as well as under the successor t. It then follows by freeness of O that $A = O$. For small sups, assume $\{x_i\}$ is a small collection of elements of A, and consider $x = \bigvee x_i$. Since $x_i \leq x$, also $t(x_i) \leq t(x)$ for each i. Thus, since $x_i \leq t(x_i)$ by assumption, also $x_i \leq t(x)$. This holds for all i, so $x \leq t(x)$. This shows $x \in A$. For the successor, suppose $x \in A$, so that $x \leq t(x)$. Since t is monotone, also $t(x) \leq tt(x)$, i.e. $t(x) \in A$. This shows that A is closed under small sups and successor, as required.

In the next section, we will consider the *free* ZF-algebra equipped with an inflationary successor. This algebra is in general different from the algebra (O, t).

Next, we derive an analog of Theorems 1.2 and 1.5 for the algebra of ordinals (O, t). Again, this result can be viewed as a special case of Appendix A, Theorem 1.

Define a preorder $\preceq$ on $P_s(O)$, by setting for $E, F \in P_s(O)$,

$$E \preceq F \quad \text{iff} \quad \forall x \in E \ \exists y \in F \ x \leq y. \tag{1}$$

This preorder $\preceq$ defines an equivalence relation $\sim$ on $P_s(O)$ in the usual way:

$$E \sim F \quad \text{iff} \quad E \preceq F \quad \text{and} \quad F \preceq E.$$

Let

$$\mathcal{D}_s(O) := P_s(O)/\sim.$$

One should think of $\mathcal{D}_s(O)$ as the object of downwards closed subclasses of O which are generated by small sets. To suggest this, we write $\downarrow(E)$ for the equivalence class in $\mathcal{D}_s(O)$ of an element $E \in P_s(O)$.

The preorder $\preceq$ on $P_s(O)$ induces a partial order on $\mathcal{D}_s(O)$. Equipped with this order, $\mathcal{D}_s(O)$ is a sup-lattice, with smallest element $\downarrow(\emptyset)$, and binary supremum given by

$$\downarrow(E) \vee \downarrow(F) = \downarrow(E \cup F).$$

More generally, small suprema exist in $\mathcal{D}_s(O)$, and are constructed from suprema (unions) in $P_s(O)$. The proof uses the collection axiom for the class $\mathcal{S}$ of small maps.

2.2 Proposition. *The sup-lattice $\mathcal{D}_s(O)$ is $\mathcal{S}$-complete. For a small family $\mathcal{E} = \{E_i | i \in I\} \in P_s P_s(O)$ of small subsets of O, the sup in $\mathcal{D}_s(O)$ is computed as*

$$\bigvee_{i \in I} \downarrow(E_i) = \downarrow\left(\bigcup_{i \in I} E_i\right). \tag{2}$$

Proof. Consider the quotient map

$$\downarrow : P_s(O) \twoheadrightarrow \mathcal{D}_s(O).$$

By the collection axiom (A7), the induced map $P_s(\downarrow) : P_s P_s(O) \to P_s(\mathcal{D}_s(O))$ is again epi (cf. Proposition I.3.7). We claim that there exists a unique factorization, as dotted in the diagram

$$
\begin{array}{ccc}
P_s P_s(O) & \xrightarrow{\ P_s(\downarrow)\ } & P_s \mathcal{D}_s(O) \\
{\scriptstyle \cup}\big\downarrow & & \big\downarrow{\scriptstyle V} \\
P_s(O) & \xrightarrow{\ \ \downarrow\ \ } & \mathcal{D}_s(O).
\end{array}
$$

Indeed, since in a pretopos any epi is the coequalizer of its kernel pair, the existence of the dotted map follows from the fact that $\downarrow \circ \bigcup$ equalizes the kernel pair of $P_s(\downarrow)$; or, in set-theoretic notation, that for any $\mathcal{E} = \{E_i \mid i \in I\}$ and $\mathcal{F} = \{F_j \mid j \in J\}$ in $P_s P_s(O)$, if

$$\{\downarrow(E_i) \mid i \in I\} = \{\downarrow(F_j) \mid j \in J\} \quad \text{in} \quad P_s \mathcal{D}_s(O),$$

then also

$$\downarrow\left(\bigcup_{i \in I} E_i\right) = \downarrow\left(\bigcup_{j \in J} F_j\right) \quad \text{in} \quad \mathcal{D}_s(O).$$

Finally, one readily checks that this map $\bigvee : P_s \mathcal{D}_s(O) \to \mathcal{D}_s(O)$ indeed defines suprema for $\mathcal{D}_s(O)$.

Now define a successor $\bar{t}$ on $\mathcal{D}_s(O)$ by the formula

$$\bar{t}(\downarrow E) = \downarrow\{\bigvee_{x \in E} t(x)\}. \tag{3}$$

This successor $\bar{t}$ is well-defined on equivalence classes; for t is order preserving, so if $E \preceq F$ in $P_s(O)$ then also $\bigvee_{x \in E} t(x) \leq \bigvee_{y \in F} t(y)$ in O, whence $\{\bigvee_{x \in E} t(x)\} \preceq \{\bigvee_{y \in F} t(y)\}$ in $\mathcal{D}_s(O)$. The same argument shows that $\bar{t} : \mathcal{D}_s(O) \to \mathcal{D}_s(O)$ is monotone.

2.3 Theorem. *The map* $\theta : (\mathcal{D}_s(O), \bar{t}) \to (O, t)$ *defined by*

$$\theta(\downarrow E) = \bigvee_{x \in E} t(x)$$

is an isomorphism of ZF-*algebras.*

Proof. It is clear that the map θ is well-defined on equivalence classes, as well as order preserving. The map θ also preserves small suprema, since by the preceding proposition these are computed as unions in $\mathcal{D}_s(O)$. Furthermore, for $E \in P_s(O)$,

$$\theta(\bar{t}(\downarrow E)) = t(\bigvee_{x \in E} t(x))$$

$$= t(\theta(\downarrow E)).$$

This shows that $\theta : \mathcal{D}_s(O) \to O$ is a homomorphism of ZF-algebras.

On the other hand, the successor $\bar{t}$ on $\mathcal{D}_s(O)$ is monotone, and (O, t) is the free algebra with a monotone successor, so there is a unique homomorphism

$$\eta : (O, t) \to (\mathcal{D}_s(O), \bar{t}).$$

By freeness, $\theta \circ \eta$ is the identity on O. This means that for any $x \in O$ and any $E_x \in P_s(O)$ representing $\eta(x)$ (that is, $\eta(x) = \downarrow(E_x)$), one has

$$x = \theta\eta(x) = \bigvee_{y \in E_x} t(y). \tag{4}$$

Now consider an arbitrary $E \in P_s(O)$ and its associated equivalence class $\downarrow(E) \in \mathcal{D}_s(O)$. The map $\eta : O \to \mathcal{D}_s(O)$ gives a small subset

$$A = \{\eta(x) \mid x \in E\} \subseteq \mathcal{D}_s(O).$$

By collection, the map $P_s(\downarrow) : P_s P_s(O) \to P_s \mathcal{D}_s(O)$ is epi, so there is a small family $\mathcal{E} = \{E_i \mid i \in I\}$ of small subsets of O, such that

$$A = \{\downarrow(E_i) \mid i \in I\}. \tag{5}$$

But then

$$
\begin{aligned}
\eta\theta(\downarrow E) &= \eta(\textstyle\bigvee_{x \in E} t(x)) \\[4pt]
&= \textstyle\bigvee_{x \in E} \eta\, t(x) \\[4pt]
&= \textstyle\bigvee_{x \in E} \bar{t}\eta(x) \\[4pt]
&= \textstyle\bigvee_{i \in I} \bar{t}(\downarrow E_i) && \text{(by (5))} \\[4pt]
&= \textstyle\bigvee_{i \in I} \downarrow\{\bigvee_{y \in E_i} t(y)\} \\[4pt]
&= \downarrow(\textstyle\bigcup_{i \in I}\{\bigvee_{y \in E_i} t(y)\}) && \text{(by (2))} \\[4pt]
&= \downarrow\{\textstyle\bigvee_{y \in E_i} t(y) \mid i \in I\}.
\end{aligned}
$$

But each E_i represents some $\eta(x)$ where $x \in E$, so by (4) we have for such an x that $x = \bigvee_{y \in E_i} t(y)$. Since every $x \in E$ is so represented by some E_i, it follows that

$$\{\textstyle\bigvee_{y \in E_i} t(y) \mid i \in I\} = \{x \mid x \in E\} = E.$$

With the previous computation, this yields

$$\eta\theta(\downarrow E) = \downarrow(E).$$

This proves that η and θ are mutually inverse isomorphisms, and the theorem is established.

2.4 Corollary. *The algebra (O, t) of ordinal numbers has the following properties:*

(i) *For all $x, y \in O$, if $t(x) \leq t(y)$ then $x \leq y$.*

(ii) *For any $x \in O$, the family $\{y \in O | ty \leq x\}$ is generated by a small set, and $x = \bigvee_{ty \leq x} ty$.*

(iii) *(Irreducibility of successors) For any $y \in O$ and any small $E \subseteq O$, if $t(y) \leq \bigvee_{x \in E} x$ then $\exists x \in E$ $(t(y) \leq x)$.*

Proof. (i) For $x, y \in O$, let E_x and E_y represent $\eta(x)$ and $\eta(y)$, and write, as in (4),

$$x = \bigvee_{a \in E_x} t(a), \quad y = \bigvee_{b \in E_y} t(b).$$

Thus

$$t(x) = \theta \bar{t} \eta(x) = \theta(\downarrow \{\bigvee_{a \in E_x} t(a)\}) = \theta(\downarrow \{x\}),$$

and similarly

$$t(y) = \theta(\downarrow \{y\}). \tag{6}$$

But then

$$t(x) \leq t(y) \quad \text{iff} \quad \theta(\downarrow \{x\}) \leq \theta(\downarrow \{y\})$$

$$\text{iff} \quad \downarrow \{x\} \leq \downarrow \{y\}$$

$$\text{iff} \quad \{x\} \preceq \{y\}$$

$$\text{iff} \quad x \leq y.$$

(ii) For $x \in O$ let E_x represent $\eta(x)$ so that $x = \bigvee_{a \in E_x} t(a)$, as before. Then for any $y \in O$, (6) yields

$$t(y) \leq x \quad \text{iff} \quad \theta(\downarrow \{y\}) \leq x$$

$$\text{iff} \quad \downarrow \{y\} \leq \eta(x) = \downarrow (E_x)$$

$$\text{iff} \quad \exists a \in E_x \ y \leq a.$$

Thus $\{y | t(y) \leq x\}$ is generated by the small set E_x. Furthermore, clearly $x = \bigvee_{t(y) \leq x} t(y)$, since $x = \bigvee_{a \in E_x} t(a)$.

(iii) For the small set $E \subseteq O$, let $\{E_i | i \in I\}$ be a small collection of small subsets of O with the property that

$$\{\eta(x) \mid x \in E\} = \{\downarrow(E_i) \mid i \in I\},$$

as in (5) above. Then, again using (6),

$$t(y) \leq \bigvee_{x \in E} x \quad \text{iff} \quad \theta(\downarrow\{y\}) \leq \bigvee_{x \in E} x$$

$$\text{iff} \quad \downarrow\{y\} \leq \eta(\bigvee_{x \in E} x) = \bigvee_{x \in E} \eta(x)$$

$$\text{iff} \quad \downarrow\{y\} \leq \downarrow(\bigcup_{i \in I} E_i)$$

$$\text{iff} \quad \exists i \in I \; \exists w \in E_i \; y \leq w.$$

But this E_i represents $\eta(x)$ for some $x \in E$, so for this x we find $ty \leq tw \leq \bigvee_{a \in E_x} t(a) = x$.

This completes the proof of Corollary 2.4.

Since the algebra (V, s) considered in the previous section is initial, there is a unique homomorphism of ZF-algebras $(V, s) \rightarrow (O, t)$, which we denote

$$rank : V \rightarrow O. \tag{7}$$

In other words, this rank function satisfies the usual identities

$$rank(sx) = t(rank(x)), \quad rank(\bigvee x_i) = \bigvee rank(x_i)$$

for successor and small sups, and is uniquely determined by these identities.

We now show that the algebra of ordinals can be used to build up the free ZF-algebra V as a cumulative hierarchy indexed by the ordinals, as expressed by Theorem 2.7 below. For this, we shall make the additional assumption that the class $\mathcal{S}$ of small maps $\mathcal{C}$ satisfies the following "power-set axiom" (cf. Remarks I.3.2, I.3.4):

(S3) ("Power-set Axiom") For any small map $X \rightarrow B$ in the category $\mathcal{C}$, the power-object $P_s(X \rightarrow B)$ is a small object of $\mathcal{C}/B$.

Recall that, given the earlier axioms for small maps, this axiom is equivalent to the condition that for each object B of $\mathcal{C}$, the category $\mathcal{S}/B$ of small objects over B is an elementary topos.

2.5 Proposition. *Suppose the class of small maps S satisfies the power-set axiom (S3). Then there exists an "internal" power-set operation*

$$p : V \to V$$

with the property that for all $x, y \in V$,

$$y \leq x \quad iff \quad sy \leq p(x).$$

Proof. We use the mutually inverse isomorphisms

$$V \underset{r}{\overset{i}{\rightleftarrows}} P_s(V)$$

of Theorem 1.2. Thus $i(x) = \{y | sy \leq x\}$ while $r(E) = \bigvee_{x \in E} sx$. Define, for $x \in V$,

$$p(x) = \bigvee_{E \in P_s(ix)} sr(E).$$

Here we view $i(x)$ as a small subobject of V, so that any $E \in P_s(ix)$ can be viewed as an element of $P_s(V)$, and $r(E)$ thus makes sense. The supremum occurring in the definition of $p(x)$ exists, since by the power-set axiom $P_s(ix)$ is small. Now for any $y \in V$,

$$s(y) \leq p(x) \quad \text{iff} \quad \exists E \in P_s(ix) \;\; s(y) \leq sr(E) \quad \text{(by 1.4(iii))}$$

$$\text{iff} \quad \exists E \in P_s(ix) \;\; y = r(E) \qquad \text{(by 1.4(ii))}$$

$$\text{iff} \quad \exists E \in P_s(ix) \;\; iy = E$$

$$\text{iff} \quad iy \leq ix$$

$$\text{iff} \quad y \leq x.$$

This proves Proposition 2.5.

2.6 Corollary. *For any $x \in V$, its down segment $\downarrow(x) = \{y \in V | y \leq x\}$ is a small set. Or, in other words, (S3) implies that the second projection $\{(y, x) | y \leq x\} \to V$ belongs to S.*

Proof. Immediate from 2.5 and 1.4(ii).

2.7 Theorem. *Suppose the class S of small maps satisfies the powerset axiom (S3). Then the map rank $: V \to O$ has a right adjoint $V_{(-)} : O \to V$,* *i.e.*

(i) $x \leq V_\alpha$ *iff* $rank(x) \leq \alpha$ *(any* $x \in V, \alpha \in O$),

   having the following additional properties:

(ii) $V_\alpha = \bigvee_{i \in I} V_{\alpha_i}$, *for any small sup* $\alpha = \bigvee \alpha_i$ *in* O,

(iii) $V_{t\alpha} = p(V_\alpha)$, *for any* $\alpha \in O$,

(iv) $rank(V_\alpha) = \alpha$, *for any* $\alpha \in O$.

Proof. Consider the powerset operation $p : V \to V$ constructed in Proposition 2.5. From the property

$$y \leq x \quad \text{iff} \quad sy \leq p(x)$$

stated there, together with Corollary 1.4(ii), it is clear that p is monotone. Thus (V, p) is a ZF-algebra with a monotone successor. Since (O, t) is the free one, there exists a unique homomorphism of ZF-algebras $(O, t) \to (V, p)$, which we denote

$$V_{(-)} : O \to V.$$

The identities (ii) and (iii) of the theorem express that $V_{(-)}$ is a homomorphism of ZF-algebras. Furthermore, again by freeness of (O, t), the composite homomorphism $rank \circ V_{(-)} : O \to O$ must be the identity, thus proving property (iv). It remains to prove the adjointness. Since we already showed that $V_{(-)}$ is right inverse to $rank$, it is in fact enough to show that $x \leq V_{rank(x)}$ for any $x \in V$. Let $A = \{x \in V | x \leq V_{rank(x)}\}$. Then A is closed under small sups, since the operations $rank$ and $V_{(-)}$ preserve small sups. Furthermore, A is closed under the successor $s : V \to V$, since if $x \leq V_{rank(x)}$ then also

$$
\begin{aligned}
s(x) \;\; &\leq \;\; p(x) && \text{(by Proposition 2.5)} \\
&\leq \;\; p(V_{rank(x)}) && (p \text{ is monotone}) \\
&= \;\; V_{t(rank(x))} && \text{(by part (iii))} \\
&= \;\; V_{rank(sx)}.
\end{aligned}
$$

Since (V, s) is initial, it follows that $A = V$. Thus $x \leq V_{rank(x)}$ for all $x \in V$, as required.

This completes the proof of the theorem.

2.8 Remark. Consider the composite $V_{(-)} \circ rank : V \to V$. It follows from Theorem 2.7 that this map is idempotent, and that $O = \{x \in V \,|\, V_{rank(x)} = x\}$. This argument can also be reversed, to provide a construction of O from V. Indeed, independent of O, freeness of (V, s) gives a unique homomorphism $\pi : (V, s) \to (V, p)$. It is not difficult to show that $\{x \in V \,|\, \pi^2(x) = \pi(x)\}$ is closed under small sups and successor, so that π must be idempotent. Thus there is a coequalizer for the image of π,

$$V \underset{id}{\overset{\pi}{\rightrightarrows}} V \overset{\pi}{\to} Im(\pi),$$

by means of which one can show directly that $(Im(\pi), p)$ is the free initial ZF-algebra with monotone successor. (Cf. Theorem 3.1 in the next section for an analogous argument.)

§3   Von Neumann ordinals

In this section we will consider the free ZF-algebra with an inflationary successor . We will denote it by

$$(N, \bar{s}),$$

and call it the algebra of *Von Neumann ordinals* . Thus

$$x \leq \bar{s}(x)$$

for all $x \in N$, and $(N, \bar{s})$ is initial with this property. We will show that $(N, \bar{s})$ can be constructed from the algebra (V, s) considered in §1, as the subobject consisting of "hereditarily transitive sets" (Theorem 3.4 below).

For a first construction of N from V, consider the map

$$\bar{s} : V \to V \ , \ \ \bar{s}(x) = x \vee s(x).$$

Viewing $\bar{s}$ as a successor operation, this gives a new ZF-algebra $(V, \bar{s})$. Since (V, s) is the free ZF-algebra, there is thus a unique homomorphism $\rho : (V, s) \to (V, \bar{s})$. The map $\rho : V \to V$ is the unique one with the properties

$$\rho(\bigvee x_i) = \bigvee \rho(x_i) \ , \ \ \rho(s(x)) = \rho(x) \vee s(\rho(x)), \tag{1}$$

for any small supremum $\bigvee x_i$ and any $x \in V$.

3.1 Theorem. *The map ρ is idempotent, its image $\rho(V) \subseteq V$ is closed under the operation $\bar{s}$, and $(\rho(V), \bar{s})$ is (isomorphic to) the initial ZF-algebra with an inflationary successor .*

On the basis of this theorem, we will identify $(N, \bar{s})$ with $(\rho(V), \bar{s})$.

Proof. To see that $\rho^2 = \rho$, it suffices to show that $\rho^2 : V \to V$ satisfies the same defining equations (1) as ρ does. Clearly, if ρ preserves all small sups, then so does ρ^2.

Furthermore, for any $x \in V$ one has

$$\rho^2(sx) = \rho(\rho(x) \vee s\rho(x))$$

$$= \rho^2(x) \vee \rho(s\rho(x))$$

$$= \rho^2(x) \vee (\rho^2(x) \vee s\rho^2(x))$$

$$= \rho^2(x) \vee s\rho^2(x).$$

This shows $\rho^2 = \rho$.

Next, the image $\rho(V)$ is clearly closed under the operation $\bar{s}$, since $\bar{s}(\rho(x)) = \rho(x) \vee s\rho(x) = \rho(sx)$. Thus $(\rho(V), \bar{s})$ is a ZF-algebra. In order to show that this algebra is the free algebra with an inflationary successor, note first that $\rho(V)$ fits into a split coequalizer

$$V \underset{id}{\overset{\rho}{\rightrightarrows}} V \overset{\rho}{\longrightarrow} \rho(V),$$

as for any idempotent map. Now suppose (L, t) is an arbitrary ZF-algebra with a successor t having the property that $x \leq t(x)$ for all $x \in L$. By freeness of (V, s), there is a unique map $\varphi : V \to L$ which preserves small sups and satisfies $\varphi(sx) = t(\varphi(x))$. We claim that $\varphi = \varphi\rho$. Clearly $\varphi\rho$ preserves small sups. Furthermore

$$\varphi\rho(sx) = \varphi(\rho(x) \vee s\rho(x))$$

$$= \varphi\rho(x) \vee \varphi s\rho(x)$$

$$= \varphi\rho(x) \vee t(\varphi\rho(x))$$

$$= t(\varphi\rho(x)),$$

the last since t is inflationary. Thus $\varphi\rho$ preserves the successor. This shows that $\varphi\rho$ satisfies the conditions which uniquely determine φ, hence $\varphi = \varphi\rho$. By the coequalizer above, φ thus factors uniquely as $\varphi = \psi \circ \rho$, where the map $\psi : (\rho(V), \bar{s}) \to (L, t)$ preserves small sups and successor since φ and $\rho : (V, s) \to (\rho(V), \bar{s})$ do. Moreover, ψ is clearly unique with this property. This completes the proof.

3.2 Corollary. *In the algebra* $(N, \bar{s})$ *of Von Neumann ordinals, successors are irreducible : for any* $y \in N$ *and any small* $E \subseteq N$,

$$\bar{s}(y) \le \bigvee_{x \in E} x \quad \Rightarrow \quad \exists x \in E \ (\bar{s}(y) \le x).$$

Proof. By Theorem 3.1, it suffices to show that for any $y \in \rho(V)$ and any small family $\{x_i : i \in I\}$ of elements of $\rho(V)$, if $\bar{s}(y) \le \bigvee x_i$ then $\bar{s}(y) \le x_i$ for some i. Suppose $\bar{s}(y) \le \bigvee x_i$. Since $s(y) \le \bar{s}(y)$ and successors are irreducible in V (Corollary 1.4 (iii)), we find $s(y) \le x_i$ for some i. But $y, x_i \in \rho(V)$, so these are fixed points of ρ. Thus

$$y \vee s(y) = \rho(sy) \le \rho(x_i) = x_i,$$

or $\bar{s}(y) \le x_i$, as required.

Note that since ρ is idempotent, its image $N = \rho(V)$ coincides with the equalizer of ρ and $id : V \rightrightarrows V$. Thus

$$N = \{x \in V \mid \rho(x) = x\}.$$

We will now show that N can be described by the familar definition of ordinal numbers as "transitive sets of transitive sets" . To this end, define an operation

$$\cup : V \to V$$

for "union" by

$$\cup x = \bigvee_{s(y) \le x} y. \tag{2}$$

Notice that this is well-defined, since by Corollary 1.4(ii) the supremum in (2) is small. Also notice that by part (i) of that corollary, the identity

$$\cup s(x) = x \tag{3}$$

holds. Now define the following two subobjects of V:

$$T = \{x \in V \mid \cup x \le x\},$$

$$T^{(2)} = \{x \in T \mid \forall y \in V(sy \le x \ \Rightarrow \ y \in T)\}.$$

3.3 Lemma. *Both $T \subseteq V$ and $T^{(2)} \subseteq V$ are closed under small suprema as well as under the operation $\bar{s} : V \to V$, $\bar{s}(x) = x \vee s(x)$.*

Proof. By irreducibility of successors in V (Corollary 1.4(iii)), we have

$$\cup(\textstyle\bigvee_{i \in I} x_i) \;=\; \textstyle\bigvee\{y \mid s(y) \leq \textstyle\bigvee_{i \in I} x_i\}$$

$$= \textstyle\bigvee\{y \mid \exists i \in I \;\; s(y) \leq x_i\}$$

$$= \textstyle\bigvee_{i \in I} \textstyle\bigvee\{y \mid s(y) \leq x_i\}$$

$$= \textstyle\bigvee_{i \in I}(\cup x_i).$$

Thus T is closed under small sups, because these are preserved by the union operation. Furthermore,

$$\cup\, \bar{s}(x) \;=\; \cup\,(x \vee s(x))$$

$$= \cup x \vee \cup s(x)$$

$$= \cup x \vee x,$$

the latter by (3) above. Thus if $\cup x \leq x$ then $\cup \bar{s}(x) = x \leq \bar{s}(x)$, showing that T is closed under the operation $\bar{s}$.

For $T^{(2)}$, it follows again from irreducibility of successors in V that $T^{(2)}$ is closed under small sups. For closure under $\bar{s}$, suppose $x \in T^{(2)}$ and consider $y \in V$ with $s(y) \leq \bar{s}(x) = x \vee s(x)$. Then either $s(y) \leq x$ in which case $y \in T$ since $x \in T^{(2)}$, or $y = x$ in which case clearly $y \in T$. Thus $\bar{s}(x) \in T^{(2)}$.

This proves the lemma.

3.4 Theorem. $N = T^{(2)}$. *Thus $(T^{(2)}, \bar{s})$ is the initial ZF-algebra with an inflationary successor.*

Proof. Recall that for the operation $\rho : V \to V$ we have made the identification

$$N = \rho(V) = \{x \in V \mid \rho(x) = x\}.$$

To show that $N = \rho(V) \subseteq T^{(2)}$, consider the subobject C of V defined by $C = \{x \in V \mid \rho(x) \in T^{(2)}\}$. Since $T^{(2)}$ is closed under small suprema and under $\bar{s}$, it is clear from the definition (1) of ρ that C is closed under small sups and $s : V \to V$. Since (V, s) is initial, it follows that $C = V$, whence $N \subseteq T^{(2)}$.

For the converse inclusion $T^{(2)} \subseteq N$, we prove that $\rho(x) = x$ for all $x \in T^{(2)}$. To this end, define a subobject $A \subseteq V$ by

$$A = \{x \in V \mid (x \in T^{(2)} \Rightarrow \rho(x) = x) \ \& \ \forall y \in T^{(2)}(sy \leq x \Rightarrow \rho(y) = y)\}.$$

To show $A = V$, it suffices to show that A is closed under small suprema and under the successor operation $s : V \to V$. For small sups, suppose $x = \bigvee x_i$ where each $x_i \in A$. Then, by irreducibility of successors, any y with $sy \leq x$ also satisfies $sy \leq x_i$ for some x_i, so x surely satisfies the second requirement in the definition of A. For the first requirement, assume $x \in T^{(2)}$, and write (cf. Corollary 1.4)

$$x = \bigvee\nolimits_{sy \in x} sy = \bigvee\nolimits_i \bigvee\nolimits_{sy \leq x_i} sy.$$

Now $sy \leq x_i$ implies $\rho y = y$, since $x_i \in A$ by assumption. Thus

$$
\begin{aligned}
\rho x &= \textstyle\bigvee_i \bigvee_{sy \leq x_i} \ \rho(sy) \\[4pt]
&= \textstyle\bigvee_i \bigvee_{sy \leq x_i} \ [\rho(y) \vee s\rho(y)] \\[4pt]
&= \textstyle\bigvee_i \bigvee_{sy \leq x_i} \ [y \vee s(y)] \\[4pt]
&= \textstyle\bigvee_{sy \leq x} y \ \vee \ \bigvee_{sy \leq x} s(y) \\[4pt]
&= \cup x \ \vee \ x.
\end{aligned}
$$

But $x \in T^{(2)}$ by assumption, so $\cup x \vee x = x$, whence $\rho x = x$ as desired. This shows that A is closed under small sups.

For successors, suppose $x = sz$ with $z \in A$. We show $x \in A$. First, for any $y \in V$, Corollary 1.4 gives that $sy \leq x$ implies $y = z$. So clearly x satisfies the second requirement in the definition of A. For the first requirement, suppose $x \in T^{(2)}$. Then also $z \in T^{(2)}$, since $T^{(2)}$ is itself "transitive", in the sense that $a \varepsilon b \in T^{(2)}$ implies $a \in T^{(2)}$. Hence since $z \in A$ by assumption, also $\rho(z) = z$. Therefore $\rho(x) = \rho(z) \vee s\rho(z) = z \vee s(z) = z \vee x$. But $x \in T$ and $z \varepsilon x$, so $z \leq x$. Thus $\rho(x) = z \vee x = x$. This proves $x \in A$.

This completes the proof of the theorem.

As a consequence, one obtains an analog of Theorem 1.2 for the algebra N of Von Neumann ordinals. Let

$$\mathcal{D}(N) = \{E \in P_s(N) \mid \forall x \in E \ \forall y \in N(sy \leq x \Rightarrow y \in E)\}.$$

Thus $\mathcal{D}(N)$ is the poset of small down-segments in N, for the "strict" order defined on N by $y < x$ iff $sy \leq x$.

3.5 Corollary. *The isomorphism* $V \cong P_s(V)$ *restricts to an isomorphism*

$$N \cong \mathcal{D}(N).$$

Proof. The isomorphism $i : V \to P_s(V)$ of Theorem 1.2, defined by $i(x) = \{y \in V \mid sy \leq x\}$, clearly maps N into $\mathcal{D}(N)$, because for $x \in N$ and $sy \leq x$ also $y \in N$. In the other direction, $\mathcal{D}(N) \subseteq P_s(V)$, and we will show that for any $E \in \mathcal{D}(N)$ the isomorphism $r : P_s(V) \to V$ sends E to an element $r(E)$ of N. Indeed, $x \varepsilon r(E)$ iff $x \in E$, so clearly $E \subseteq N$ gives $\forall x \varepsilon r(E)(x \in T)$. Furthermore, $y \varepsilon x \varepsilon r(E)$ gives $y \varepsilon x \in E$, hence $y \in E$ since E is downwards closed; thus $y \varepsilon r(E)$. This shows that $r(E) \in T$. In other words, $r(E) \in T^{(2)} = N$. We have proved that the mutually inverse isomorphisms $i : V \to P_s(V)$ and $r : P_s(V) \to V$ of Theorem 1.2 restrict to maps $N \to P_s(N)$ and $P_s(N) \to N$, thus proving the corollary.

3.6 Remark. We consider the possible monotoneity of the successor operation $\bar{s} : N \to N$ of the algebra $N = \rho(V) = T^{(2)}$ of Von Neumann ordinals. For $x \leq y$ in N, the inequality $\bar{s}x \leq \bar{s}y$ means that $sx \leq x \lor s(x) = \bar{s}x \leq y \lor s(y)$, hence $x \varepsilon y$ or $x = y$, by irreducibility of successors. So $\bar{s} : N \to N$ is monotone iff

$$x \leq y \;\Rightarrow\; x \varepsilon y \text{ or } x = y \quad (\text{all } x, y \in N \subseteq V). \tag{4}$$

It follows that if $\bar{s} : N \to N$ is monotone then any small mono $U \rightarrowtail X$ has a complement. Indeed, it suffices to prove this for $X = 1$ (replace $\mathcal{C}$ by $\mathcal{C}/X$). We use the internal logic, and let $U \subseteq 1 = \{*\}$. Note that $\emptyset \in V$ (empty sup), hence $\emptyset \in N$ since $N \subseteq V$ is closed under small sups. Then also $\bar{s}\emptyset = \emptyset \lor s\emptyset = s\emptyset \in N$. Let

$$\bar{U} = \bigvee \{s\emptyset \mid * \in U\}.$$

Thus $\bar{U} \in N$ whenever U is small, again since N is closed under small sups. But clearly $\bar{U} \leq s\emptyset$, so (4) would imply that $\bar{U} \varepsilon s\emptyset$ or $\bar{U} = s\emptyset$. In the first case $\bar{U} = \emptyset$, hence not $* \in U$, i.e. $U = 0$. In the second case, $\emptyset \varepsilon \bar{U}$ hence $* \in U$, i.e. $U = 1$.

In particular, if the class $\mathcal{S}$ of small maps in $\mathcal{C}$ satisfies the separation axiom, stating that every mono is small (see Theorem 4.6 below and §5 of this chapter), the successor $\bar{s} : N \to N$ is monotone iff every subobject in $\mathcal{C}$ is complemented. Or in other words, for such a class $\mathcal{S}$,

$$N = O \quad \text{iff} \quad \mathcal{C} \text{ is Boolean.}$$

3.7 Remark. By Proposition 2.1 and freeness of the algebra $(N, \bar{s})$, there is a unique homomorphism

$$h : (N, \bar{s}) \to (O, t),$$

showing that O is a quotient algebra of N. Now suppose that the class S of small maps satisfies the power-set axiom (S3), discussed in the previous section. Then we can define a monotone successor $\sigma : N \to N$, similar to the "power-set" successor $p : V \to V$, by

$$\sigma(x) = \bigvee\nolimits_{y \leq x} \bar{s}(y) \quad (x, y \in N).$$

Let $i : (O, t) \to (N, \sigma)$ be the unique homomorphism of ZF-algebras for this new successor σ. The composite map $h \circ i : O \to O$ preserves small sups, and by monotonicity of the successor t, it also preserves successors, since

$$
\begin{aligned}
hi(tx) &= h(\sigma i(x)) \\[2mm]
&= \bigvee\nolimits_{y \leq i(x)} h(\bar{s}y) \\[2mm]
&= \bigvee\nolimits_{y \leq i(x)} t(h(y)) \\[2mm]
&= t(hi(x)).
\end{aligned}
$$

By freeness of (O, t), it follows that $h \circ i = id$. Thus O is a retract of N. In particular, by Theorem 3.4, the algebra O can be taken to consist of transitive sets of transitive sets. Is it possible to describe the algebra (O, t) directly in terms of an explicit "set-theoretic" property of such hereditarily transitive sets, analogous to the description of the ordinal algebra $(N, \bar{s})$ in Theorem 3.4?

We conclude this section with some remarks on the subobject $T \subseteq V$ defined above (cf. Lemma 3.3). There is always an operation of "*transitive closure*"

$$\tau : V \to V,$$

preserving small sups and satisfying

$$\tau(sx) = \tau(x) \vee s(x).$$

To construct τ, first define a successor $k : V \times V \to V \times V$ by $k(x, y) = (s(x), y \vee s(x))$. This makes $V \times V$ into a ZF-algebra. By freeness of V, there is a unique map

$$(\lambda, \tau) : V \to V \times V$$

which preserves small sups and satisfies $(\lambda, \tau)(sx) = k((\lambda, \tau)(x))$. Thus both λ and τ are sup-preserving maps $V \to V$, and $\lambda(sx) = s\lambda(x)$ while $\tau(sx) = \tau(x) \vee s(\lambda x)$. By freeness of V, the map λ must be the identity; hence $\tau(sx) = \tau(x) \vee s(x)$, as desired.

3.8 Lemma. *The map* $\tau : V \to V$ *is idempotent and satisfies* $x \leq \tau(x)$.

Proof. One easily shows that the subobject $B \subseteq V$ defined by $B = \{x \in V \mid x \leq \tau(x) = \tau^2(x)\}$ is closed under successors and small sups. Hence $B = V$, since (V, s) is free.

3.9 Proposition. $T = \{x \in V \mid \tau(x) = x\}$.

Proof. ($\supseteq$) Let $C = \{x \in V \mid \cup x \leq \tau(x)\}$. Then C is closed under small sups, since these are preserved by both $\cup$ and τ. C is also closed under the successor s, since by (3), $x \in C$ gives

$$\cup s(x) = x \ \leq \ \tau(x) \ \leq \ \tau(sx),$$

hence $s(x) \in C$. Thus $C = V$, and hence $\cup x \leq x$ whenever $\tau(x) = x$.

($\subseteq$) For this inclusion, we use the natural numbers object $\mathbf{N}$ of the ambient category $\mathcal{C}$, and the map

$$\mathbf{N} \times V \to V \ , \ \ (n, x) \mapsto \cup^{(n)} x = \cup \cdots \cup x \ \ (n \text{ times}).$$

By induction on $n \in \mathbf{N}$, one proves that for all $x \in T$

$$\forall n \in \mathbf{N}(\cup^{(n)} \ x \leq x).$$

Thus, to show $\tau(x) \leq x$ for any $x \in T$, it suffices to prove that for any $x \in V$:

$$\forall y \in V : (y \varepsilon \tau(x) \Rightarrow \exists n \in \mathbf{N} \ y \varepsilon \cup^{(n)} \ x). \tag{5}$$

To this end, let

$$A = \{x \in V \mid \forall y \varepsilon \tau(x) \ \exists n \in \mathbf{N} \ y \varepsilon \cup^{(n)} \ x\}.$$

Then clearly A is closed under small sups, since these are preserved by τ. To show that A is also closed under successors, suppose $x \in A$, and consider any $y \in V$ with $y \varepsilon \tau(sx) = \tau(x) \vee s(x)$. If $y \varepsilon \tau(x)$, then $y \varepsilon \cup^{(n)} x$ since $x \in A$ by assumption; but $\cup s(x) = x$, so $y \varepsilon \cup^{(n)} x = \cup^{(n+1)} s(x)$. And if $y \varepsilon s(x)$ then $y \varepsilon \cup^{(0)} s(x)$, so again $y \varepsilon \cup^{(n)} s(x)$ for some n. This proves that $s(x) \in A$, so that A is closed under successors as well. Thus $A = V$, proving (5) as required.

(This proof of the inclusion "$\subseteq$" is due to J. van Oosten.)

§4   The Tarski fixed point theorem

In this section we introduce yet another type of ordinal numbers, suitable for proving a "constructive" version of Tarski's fixed point theorem in the general context of a Heyting pretopos C with a class of small maps S.

Recall that a subset D of a partially ordered set P is *directed* if D is inhabited (i.e., the unique arrow $D \to 1$ is epi), and for any $d, d' \in D$ there exists a $d'' \in D$ with $d \le d''$ and $d' \le d''$. If only this last condition is satisfied, we call D *weakly directed*. Notice that in a poset P, all sups of weakly directed subsets exist iff P has a smallest element 0 and all directed sups exist.

In this section, we wish to consider the poset T in our ambient category C which is free with the following properties: T has all small directed suprema, a successor operation $r : T \to T$ which is monotone (i.e. $x \le y$ implies $rx \le ry$), and a constant $0 \in T$ with $0 \le r(0)$.

Observe that this definition differs from that of the algebras of ordinals (O, t) and $(N, \bar{s})$, in that T is not by definition a ZF-algebra, since a priori T has *directed* small suprema only. However, we will prove in this section that (T, r) is in fact a ZF-algebra, and we will refer to it as the algebra of *Tarski ordinals* .

First, we note that $0 \in T$ is the smallest element, and that r is inflationary:

4.1. Lemma. *For all $x \in T$ the inequalities*

$$0 \le x \le r(x)$$

hold.

Proof. Let $A = \{x \in T \mid 0 \le x \le r(x)\}$. To prove that $A = T$, it suffices by freeness of T to show that A contains 0 and is closed under small directed suprema and under the successor operation r. Now clearly $0 \in A$. For directed sups, suppose $x = \bigvee_{i \in I} x_i$ is a directed supremum where each $x_i \in A$. Since the index set I must then be inhabited, and $0 \le x_i$ for each i, it follows that $0 \le \bigvee x_i = x$. Also,

$$x = \bigvee x_i \le \bigvee r(x_i) \le r(\bigvee x_i) = r(x),$$

where the first inequality holds since we assume $x_i \in A$, while the second one holds since r is monotone. This proves $x \in A$. Thus A is closed under directed sups. The proof that A is closed under the operation r is even easier, and omitted. Thus $A = T$, as required.

By this lemma and the remarks preceding its statement, T is also free with small *weakly directed* sups and a monotone successor r.

Next, we show that T is isomorphic to a poset of suitable subsets of T, analogous to earlier results for the algebra O of ordinal numbers (cf. Proposition 2.2 and Theorem 2.3). To this end, and similarly to the treatment in §2, let

$$\mathcal{D}_{sd}(T)$$

be the collection of down segments in T which are generated by small weakly directed sets $E \subseteq T$, and denote such a down segment by $\downarrow(E)$. Exactly as for O in §2, the object $\mathcal{D}_{sd}(T)$ inherits a partial order from T, a bottom element $0 = \downarrow(\emptyset)$, and a successor $a : \mathcal{D}_{sd}(T) \to \mathcal{D}_{sd}(T)$ defined by

$$a(\downarrow E) = \downarrow\{\textstyle\bigvee_{x \in E} r(x)\}.$$

Also, $\mathcal{D}_{sd}(T)$ has weakly directed sups, given by unions:

$$\textstyle\bigvee_i \downarrow(E_i) \;=\; \downarrow(\bigcup E_i),$$

for $\{\downarrow(E_i) \mid i \in I\}$ small and weakly directed. Analogous to the maps η, θ for O in §2, there are maps

$$T \underset{\theta}{\overset{\eta}{\rightleftarrows}} \mathcal{D}_{sd}(T).$$

The map η is given by freeness of T, and is the unique one which preserves successor and weakly directed sups. The map θ is explicitly defined, as

$$\theta(\downarrow E) \;=\; \textstyle\bigvee_{x \in E} r(x).$$

4.2 Theorem. *These maps η and θ are mutually inverse isomorphisms* $T \cong \mathcal{D}_{sd}(T)$.

Proof. Note first that θ preserves the successor as well as weakly directed sups, so that $\theta \circ \eta = 1$. In particular, if $x \in T$ and E_x represents $\eta(x)$ (i.e., E_x is a small weakly directed subset of T and $\eta(x) = \downarrow(E_x)$, as in §2), then

$$x = \theta\, \eta(x) \;=\; \textstyle\bigvee_{y \in E_x} r(y). \tag{1}$$

Exactly as in the proof of 2.3 for O, we can then use the collection axiom to show that $\eta \circ \theta = 1$: For $E \subseteq T$ weakly directed and small, first pick a small family $\{E_i \mid i \in I\}$ such that each E_i is a weakly directed small subset of T, and such that

$$\{\eta(x) \mid x \in E\} = \{\downarrow(E_i) \mid i \in I\}. \tag{2}$$

Then

$$\eta\theta(\downarrow E) = \eta(\bigvee_{x \in E} r(x)) \quad \text{(def. of } \theta)$$

$$= \bigvee_{x \in E} a\eta(x)$$

(since η preserves weakly directed sups and successor)

$$= \bigvee_{i \in I} a(\downarrow E_i) \qquad \text{(by choice of the } E_i)$$

$$= \bigvee_{i \in I} \downarrow\{\bigvee_{y \in E_i} r(y)\} \quad \text{(def. of } a)$$

$$= \bigvee_{x \in E} \downarrow\{x\} \qquad \text{(by (1) and (2))}$$

$$= \downarrow(E).$$

This completes the proof that η and θ are mutually inverse isomorphisms.

4.3 Corollary. *For the algebra (T, r) of Tarski ordinals the following properties hold:*

(i) *For all $x, y \in T$: if $r(x) \le r(y)$ then $x \le y$.*

(ii) *For any $x \in T$, the family $\{y \in T \mid r(y) \le x\}$ is generated by a small weakly directed set, and*

$$x = \bigvee_{r(y) \le x} r(y).$$

(iii) *(Irreducibility of successors) For any $y \in T$ and any small weakly directed $E \subseteq T$, if $r(y) \le \bigvee_{x \in E} x$ then $\exists x \in E(r(y) \le x)$.*

Proof. Analogous to the proof of Corollary 2.4.

We are now ready to show that (T, r) is in fact a free ZF-algebra:

4.4 Theorem. (i) *All binary suprema exist in T, and are preserved by the successor $r : T \to T$. In particular, (T, r) is a ZF-algebra.*

(ii) (T, r) *is the free ZF-algebra on a successor* r *which preserves binary suprema.*

Proof. (i) We first show that r preserves all binary sups that exist in T. Or more precisely, if $x, y \in T$ and $x \vee y$ exists then $r(x) \vee r(y)$ exists and $r(x) \vee r(y) = r(x \vee y)$. To this end, consider the map

$$\alpha : T \to \mathcal{D}_{sd}(T) \,, \ \alpha(x) = \downarrow \{x\},$$

so that $\alpha = \eta \circ r$ and $\theta \circ \alpha = r$. Since θ is an isomorphism by Theorem 4.2, it suffices to show that α preserves a binary sup whenever this sup exists. But for any small weakly directed $E \subseteq T$,

$$\alpha(x \vee y) \leq \downarrow (E) \quad \text{iff} \quad \exists u \in E \ (x \vee y \leq u)$$

$$\text{iff} \quad \exists u \in E \ (x \leq u) \text{ and } \exists v \in E \ (y \leq v)$$

$$(\text{since } E \text{ is weakly directed})$$

$$\text{iff} \quad \alpha(x) \leq \downarrow (E) \text{ and } \alpha(y) \leq \downarrow (E).$$

Thus $\alpha(x \vee y) = \alpha(x) \vee \alpha(y)$, as claimed.

Having established this, we show that all binary sups in T exist. Write $A_x = \{y \in T \mid x \vee y \text{ exists}\}$ and $A = \{x \in T \mid A_x = T\}$. By freeness of T it suffices to show that $0 \in A$ and that A is closed under successor and small directed sups. Clearly $0 \in A$ since $0 \vee y = y$ for all y, by Lemma 4.1. Next, suppose that $x \in A$. To prove that $r(x) \in A$, or $A_{r(x)} = T$, we use freeness of T again, and show that $A_{r(x)}$ is closed under the operations. First, clearly $0 \in A_{r(x)}$. Next, if $y = \bigvee y_i$ is a small directed sup where $y_i \in A_{r(x)}$, then each $r(x) \vee y_i$ exists and hence $r(x) \vee y$ also exists since $r(x) \vee y = r(x) \vee \bigvee y_i = \bigvee (r(x) \vee y_i)$. Finally, if $y \in A_{r(x)}$ then also $r(y) \in A_{r(x)}$, since by assumption $x \in A$ so $x \vee y$ exists; thus, as shown above, $r(x) \vee r(y)$ exists (and equals $r(x \vee y)$). This shows $r(x) \in A$. It remains to be shown that A is closed under directed sups. But clearly, if $x = \bigvee x_i$ is such a sup where each $x_i \in A$, then for any $y \in T$ the supremum $x_i \vee y$ exists. Then $x \vee y$ also exists, since $x \vee y = \bigvee (x_i \vee y)$. This completes the proof of part (i) of the theorem.

(ii) As always, freeness should be interpreted with arbitrary parameters. For illustration, we will write the parameters explicitly here. So let U be any parameter object in $\mathcal{C}$, and let $P \to U$ be a ZF-algebra in $\mathcal{C}/U$ with a (fiberwise) successor operator $b : P \to P$ which preserves binary sups. By freeness of T for weakly directed sups, there is a unique map

$$\varphi : U \times T \to P, \quad \varphi(u, x) = \varphi_u(x),$$

over U, such that each φ_u preserves the successor and weakly directed small sups. It suffices to show that φ_u also preserves binary sups. To this end, define for each $x \in T$ a subobject $B_{u,x} = \{y \in T \mid \varphi_u(x \vee y) = \varphi_u(x) \vee \varphi_u(y)\}$, and let $B_u = \{x \in T \mid B_{u,x} = T\}$. Then one readily shows that $B_u = T$, by a "double induction" much as for A_x and A in part (i). We omit the details.

4.5 Remark. It follows that the algebra (T, r) of Tarski ordinals is a retract of the algebra (O, t) of ordinal numbers. Indeed, having shown that T is indeed a ZF-algebra, monotonicity of the successor r and freeness of (O, t) give a unique homomorphism of ZF-algebras $\mu : (O, t) \to (T, r)$. On the other hand, freeness of (T, r) for directed sups and a monotone successor gives a map $\nu : T \to O$. This map ν is not a ZF-algebra homomorphism, but it is the unique map which preserves the successor as well as small weakly directed sups. By freeness of T, the composite map $\mu \circ \nu$ must be the identity.

As an extension of the problem noted at the end of Remark 3.7, it is natural to ask whether there is an explicit description of T as the object of hereditarily transitive sets with a certain "set-theoretic" property.

We will come back to the inclusion of T into O when we explicitly construct T as a subobject of O in Chapter III, §5.

To conclude this section, we present the promised constructive version of Tarski's fixed point theorem. For its proof, we will assume the following additional axiom for the class of small maps $\mathcal{S}$:

(S4) ("Separation Axiom") Every monomorphism belongs to $\mathcal{S}$.

We already briefly encountered this axiom in Remark 3.6, and we will consider it again in relation to the axioms of Zermelo-Fraenkel set theory, in the next section. We point out that this separation axiom holds in most of the examples considered in Chapter IV.

Before we state the theorem, we recall that a (global) *fixed point* of an operator $\alpha : P \to P$, on a poset P in the category $\mathcal{C}$, is an arrow $p : 1 \to P$ with $\alpha \circ p = p$. It is said to be a *least* fixed point if for any arrow $q : X \to P$ in $\mathcal{C}$ with the property that $\alpha \circ q = q$, the inequality

$$(X \to 1 \xrightarrow{p} P) \leq (X \xrightarrow{q} P)$$

holds in $\mathcal{C}(X, P)$.

4.6 Theorem. *Let $\mathcal{C}$ be a Heyting pretopos, with a class of small maps $\mathcal{S}$ satisfying the separation axiom (S4). Suppose the algebra (T, r) of Tarski ordinals exists in $\mathcal{C}$. Let P be any small poset in $\mathcal{C}$ with weakly directed suprema, and let $\alpha : P \to P$ be a monotone endomorphism. Then α has a least fixed point.*

(For the existence of (T, r), see Theorem 5.4 of Chapter III.)

Proof. By freeness of (T, r), there exists a unique map $\varphi : T \to P$ which preserves small directed sups and has the property that $\varphi(r(x)) = \alpha(\varphi(x))$ for all $x \in T$. Now factor φ as $T \to Q \to P$, where $T \to Q$ is epi and $Q \to P$ is mono. Since P is assumed small, the separation axiom implies that Q is also small. Now work in the internal logic of $\mathcal{C}$, and use the collection axiom (A7) to find a small $R \subseteq T$ with $\varphi(R) = Q$. By Theorem 4.4, the small set R has a supremum $x \in T$. Let $p = \varphi(x)$. Note that p does not depend on the choice of R, since $p = \bigvee Q$. (Thus p is a global section $p : 1 \to P$.) Now $\alpha(p) = \alpha(\varphi(x)) = \varphi(r(x)) \in Q$, so $\alpha(p) \leq \bigvee Q = p$. On the other hand, $x \leq r(x)$ in T (cf. Lemma 4.1), so $p = \varphi(x) \leq \varphi(r(x)) = \alpha(\varphi(x)) = \alpha(p)$. Thus $p = \alpha(p)$, as required.

To show that p is the smallest fixed point, suppose $q \in P$ is such that $\alpha(q) = q$. It suffices to prove that $\varphi(y) \leq q$ for all $y \in T$. But, writing $A = \{y \in T \mid \varphi(y) \leq q\}$, one readily shows that A is closed under small weakly directed sups and under the successor $r : T \to T$. Thus, by freeness of (T, r), it follows that $A = T$.

This completes the proof.

Another constructive approach, leading to a double-negation variant of Tarski's fixed point theorem, is contained in Taylor(1994).

§5   Axioms for set theory

For the first order (intuitionistic) predicate calculus in a language with one unary predicate symbol A (for "atoms") and one binary symbol ε (for "membership"), we consider the following familiar axioms and axiom schemata for Zermelo-Fraenkel set theory; we use $S(x)$ ("x is a set") as an abbreviation for $\neg A(x)$.

(Z1) (Extensionality for sets) $Sx \wedge Sy \Rightarrow [x = y \Leftrightarrow \forall z(z\varepsilon x \Leftrightarrow z\varepsilon y)]$

(Z2) (Pairing) $\exists z\, [x\varepsilon z \wedge y\varepsilon z]$

(Z3) (Union) $\exists y \forall z\, [z\varepsilon y \Leftrightarrow \exists w(w\varepsilon x \wedge z\varepsilon w)]$

(Z4) (ε-Induction) $\forall x\, [\forall y\varepsilon x \varphi(y) \Rightarrow \varphi(x)] \Rightarrow \forall x \varphi(x)$

(Z5) (Power-set) $\exists y \forall z\, [z\varepsilon y \Leftrightarrow (Sz \wedge \forall w\varepsilon z(w\varepsilon x))]$

(Z6) (Separation) $\exists y\, [Sy \wedge \forall z(z\varepsilon y \Leftrightarrow z\varepsilon x \wedge \varphi)]$

(Z7) (Collection) $\forall y\varepsilon x \exists w \varphi \Rightarrow \exists z \forall y\varepsilon x \exists w\varepsilon z \varphi$

(Z8) (Infinity) $\exists x\, [\exists y(y\varepsilon x) \wedge \forall y\varepsilon x(y \cup \{y\}\varepsilon x)]$

(Z9) (Decidability between atoms and sets) $Ax \vee Sx$

(Z10) (Only sets have members) $\exists y(y\varepsilon x) \Rightarrow Sx$

(Z11) (Classical logic) $\varphi \vee \neg\varphi$

(Z12) (No atoms) $\forall x Sx$

In these axioms, we have used the standard convention of abbreviating the expressions $\forall y(y\varepsilon z \Rightarrow ...)$ and $\exists y(y\varepsilon z \wedge ...)$ to $\forall y\varepsilon z(...)$ and $\exists y\varepsilon z(...)$ respectively; also, in (Z8) we have used $\emptyset$ and $\{y\}$ as common abbreviations for the empty set and the singleton. Various combinations of these axioms give standard set theories. For example, IZFA – for "intuitionistic Zermelo-Fraenkel set theory with atoms" – is the theory axiomatized by (Z1–10). Other familiar theories are IZF , ZFA, ZF, etc. One should read these abbreviations as follows: dropping the I means adding (Z11); dropping the letter A means adding (Z12) (then (Z9), (Z10) are redundant). Thus ZF is the usual classical Zermelo-Fraenkel set theory axiomatized by (Z1–8,11,12).

We will also consider the following weakening of the separation axiom (Z6):

(Z6′) (Decidable separation)

$$\forall z(\varphi(z) \vee \neg\varphi(z)) \Rightarrow \exists y\, [Sy \wedge \forall z(z\varepsilon y \Leftrightarrow z\varepsilon x \wedge \varphi)].$$

Of course, (Z6′) is equivalent to (Z6) in the presence of the schema (Z11) for classical logic.

Now let $\mathcal{C}$ be a Heyting pretopos equipped with a class of small maps $\mathcal{S}$. Recall that, by definition, the class $\mathcal{S}$ satisfies the axioms (A1–7) for open maps, and the axioms (S1) for exponentiability and (S2) for representability. Recall also that we have already considered the following additional possible axioms for $\mathcal{S}$ (see 2.5, 4.6):

(S3) (Power-set) If $X \to B$ belongs to $\mathcal{S}$ then so does $P_s(X \to B)$.

(S4) (Separation) Every monomorphism belongs to $\mathcal{S}$.

As a last property of the class $\mathcal{S}$, we will consider in this section the axiom

(S5) (Infinity) For the natural numbers object N of the pretopos $\mathcal{C}$, the map $\mathsf{N} \to 1$ belongs to $\mathcal{S}$.

Let A be any object of $\mathcal{C}$, and let $V(A)$ be the free ZF-algebra generated by A. Construct $V'(A) = A + V(A)$ as in Remark 1.6, and define a successor

$$s : V'(A) \to V'(A) \tag{1}$$

on $V'(A)$ which extends the successor $s : V(A) \to V(A)$, by letting the restriction of s in (1) to the summand A be the composition of the map $\eta : A \to V(A)$ with the coproduct inclusion $V(A) \hookrightarrow V'(A)$. This extended successor defines an extended membership relation on $V'(A)$, by the usual formula: $x\varepsilon y$ iff $s(x) \leq y$. (But $V'(A)$ is of course *not* a ZF-algebra.) Thus $V'(A)$ provides us with an interpretation of our language with predicate symbols A and ε. In the special case where $A = 0$, one has $V'(0) = V(0) = V$, equipped with the usual successor and membership.

5.1 Proposition. *For any class of small maps $\mathcal{S}$, the axioms (Z1–4), (Z6'), (Z7), (Z9) and (Z10) hold for $V'(A)$.*

Proof. We verify the axioms in the order in which they are listed above. (Z1) By Corollary 1.7, any $y \in V(A)$ satisfies the identity

$$y = \bigvee\nolimits_{\eta(a) \leq y} \eta(a) \vee \bigvee\nolimits_{z\varepsilon y} s(z). \tag{2}$$

But $\eta(a) = s(a)$, for $a \in A$ and for the extended successor $s : V'(A) \to V'(A)$ in (1). So (2) can be rewritten as

$$y = \bigvee \{s(a) \mid a \in A, \ a\varepsilon y\} \vee \bigvee \{s(z) \mid z \in V(A), \ z\varepsilon y\}$$

$$= \bigvee \{s(w) \mid w \in A + V(A), \ w\varepsilon y\},$$

from which extensionality for $V'(A) = A + V(A)$ is evident.

(Z2) The pairing axiom holds, as witnessed by the operation

$$(x, y) \mapsto sx \vee sy$$

on $V'(A)$.

(Z3) The union axiom holds, as witnessed by the union operation $x \mapsto$

$\cup x := \bigvee_{sy \leq x} y$ discussed for V in (2) of Section 3, and the similar operation for $V(A)$.

(Z4) (ε-Induction) First suppose the formula $\varphi(x)$ contains no other free variables than x. Define two subobjects $H, C \leq 1$ in $\mathcal{C}$ by

$$H = [\![\; \forall x[\forall y \varepsilon x \varphi(y) \Rightarrow \varphi(x)] \;]\!],$$

$$C = [\![\; \forall x \varphi(x) \;]\!],$$

where x ranges over $V'(A)$. We have to show $H \subseteq C$. By passing to the slice $\mathcal{C}/H$, we may assume $H = 1$. Let $B = \{x \in V(A) \,|\, \varphi(x) \wedge \forall y \varepsilon x \varphi(y)\}$. Using $H = 1$, it is readily seen that B is closed under small suprema and the successor operation $s : V(A) \to V(A)$. Moreover, if $a \in A$ then $\eta(a) \in B$. Indeed, if $y \in V'(A)$ and $y \varepsilon \eta(a)$ then $y = a$, whence y has no elements; thus $\varphi(y)$ since $H = 1$. Since this holds for all such y, also $\varphi(\eta(a))$, again since $H = 1$. By freeness of $V(A)$ it follows that $B = V(A)$. In particular, $\forall x \in V'(A)(Sx \Rightarrow \varphi(x))$ holds. From $H = 1$ and the fact that atoms do not have elements (cf. axiom (Z10) to be verified below), also $\forall x \in V'(A)(Ax \Rightarrow \varphi(x))$. Thus $C = 1$, as required. The general case where φ is a formula $\varphi(w_1, ..., w_n, x)$ is dealt with by first passing to the appropriate slice category $\mathcal{C}/V'(A)^n$.

(Z6') (Decidable separation) We assume we are given a global section $x : 1 \to V'(A)$ and we suppose that $\varphi(z)$ has at most the variable z free. (The case where x is any generalized element and φ contains parameters is proved in exactly the same way, after replacing $\mathcal{C}$ by the appropriate slice category.) Write $1 = A_x + S_x$, where A_x and S_x are the pullbacks along x of $A \hookrightarrow V'(A)$ and $V(A) \hookrightarrow V'(A)$. Define $y : 1 \to V(A)$ by

$$y \,|\, A_x = \emptyset \; : \; A_x \to 1 \to V(A),$$

and $y \,|\, S_x : S_x \to V(A)$ constructed as follows: Working in $\mathcal{C}/S_x$, we have by Corollary 1.7 that

$$\{a \in A \,|\, \eta(a) \leq x\} \quad \text{and} \quad \{z \in V(A) \,|\, s(z) \leq x\}$$

are small subobjects of A and $V(A)$ respectively. Now by assumption on φ, we may assume that $[\![\forall z(\varphi(z) \vee \neg\varphi(z))]\!] = 1$ (possibly after passing to an appropriate slice category again). Thus the mono $\{z \in V'(A) \,|\, \varphi(z)\} \rightarrowtail V'(A)$ has a complement, hence is a small map (being a pullback of the small map $1 \rightarrowtail 1 + 1$). It follows that

$$M = \{a \in A \,|\, \varphi(a) \; \text{and} \; \eta(a) \leq x\}$$

and

$$N = \{z \in V(A) \mid \varphi(z) \text{ and } s(z) \leq x\}$$

are both small. Define

$$y = \bigvee_{a \in M} \eta(a) \vee \bigvee_{z \in N} s(z).$$

This is a global section of $V(A)$ in C/S_x, i.e. a map

$$y : S_x \to V(A)$$

in C. One readily checks, using irreducibility of successors and atoms (Corollary 1.7), that for this y the formula $S(y) \wedge \forall z (z \varepsilon y \Leftrightarrow z \varepsilon x \wedge \varphi)$ is valid in C/S_x. Since the same formula is evidently valid for $y = \emptyset$ in C/A_x and $A_x + S_x = 1$, the validity of (Z6′) has been established.

(Z7) (Collection) Here we use the collection axiom (A7) for the class S of small maps. Again, by replacing C by an appropriate slice, it suffices to consider a global section

$$x : 1 \to V'(A),$$

and a formula $\varphi(y, w)$ containing only the variables y and w free. We have to show that there is an inclusion of subobjects of 1,

$$[\![\forall y \varepsilon x \exists w \varphi]\!] \leq [\![\exists z \forall y \varepsilon x \exists w \varepsilon z \varphi]\!].$$

Again by moving into a slice of C, we may assume $[\![\forall y \varepsilon x \exists w \varphi]\!] = 1$. Define A_x and S_x exactly as in the proof of decidable separation, and work in C/A_x and C/S_x, respectively. In C/A_x, the given x is an atom, hence has no elements, so any $z : 1 \to V'(A)$ in C/A_x will witness validity in C/A_x of $\forall y \varepsilon x \exists w \varepsilon z \varphi$. On the other hand, in C/S_x the map $x : 1 \to V'(A)$ factors through the summand $V(A) \hookrightarrow V'(A)$. Thus by Corollary 1.7, both

$$D = \{a \in A \mid \eta(a) \leq x\}$$

and

$$E = \{y \in V(A) \mid s(y) \leq x\}$$

are small. Furthermore, the assumption that $[\![\forall y \varepsilon x \exists w \varphi]\!] = 1$ implies that the projection to $D + E$ is epi, as in the diagram

$$
\begin{array}{ccc}
F & \xrightarrow{\ \pi_1\ } & D + E \\
\downarrow & & \downarrow \\
V'(A) \times V'(A) & \xrightarrow{\ \pi_1\ } & A + V(A) = V'(A)
\end{array}
$$

where $F = \{(y, w) \mid y \varepsilon x \wedge \varphi(y, w)\}$. By the collection axiom for $\mathcal{S}$, there is thus a diagram of the form

$$
\begin{array}{ccc}
J & \xrightarrow{\beta} F \xrightarrow{\pi_1} & D + E \\
{\scriptstyle\alpha}\downarrow & & \downarrow \\
I & \xrightarrow{\hspace{3cm}} & 1
\end{array}
$$

where α belongs to $\mathcal{S}$ and where the map $J \to I \times (D + E)$ is epi. Define $z : I \to V(A)$ as the supremum along α of the map $J \to V(A)$ defined by

$$
J \xrightarrow{\beta} F \xrightarrow{\pi_1} A + V(A) \xrightarrow{s} V(A)
$$

(where s is the extended successor, given on the summand A by $\eta : A \to V(A)$). Then in $\mathcal{C}/I$, and for this z, the formula

$$
\forall y \varepsilon x \exists w \varepsilon z \varphi
$$

is valid.

Finally, (Z9) holds since A is a summand of $V'(A)$, and (Z10) holds by definition of ε in terms of the extended successor $s : V'(A) \to V'(A)$ (whose image is contained in the summand $V(A)$).

This completes the proof of the proposition.

The following three lemmas relate the additional properties (S3–5) for the class $\mathcal{S}$ of small maps to the remaining axioms of Zermelo-Fraenkel set theory.

5.2 Lemma. *If the class $\mathcal{S}$ satisfies the power-set axiom (S3), then (Z5) holds for $V'(A)$.*

Proof. For V, the validity of the power-set axiom (Z5) is witnessed by the operation $p : V \to V$ described in Proposition 2.5. For $V(A)$, the power-set axiom (S3) for the class $\mathcal{S}$ gives a similar operation $p : V(A) \to V(A)$ such that for any $x, z \in V(A)$,

$$
z \leq x \quad \text{iff} \quad s(z) \leq p(x).
$$

To define p, use the isomorphism $V(A) \cong P_s(A) \times P_s(V(A))$ of Theorem 1.5. Under this isomorphism, the desired map p corresponds to the operation $\tilde{p} : P_s(A) \times P_s(V(A)) \to P_s(A) \times P_s(V(A))$ defined in terms of the successor s' on $P_s(A) \times P_s V(A)$, by

$$
\tilde{p}(U, E) = \bigvee\nolimits_{V \in P_s(U), F \in P_s(E)} s'(V, F).
$$

Then the power-set axiom (Z5) for $V'(A)$ is witnessed by the extension of this map $p : V(A) \to V(A)$ to a map $p : V'(A) \to V'(A)$ which sends any atom to the bottom element $\emptyset \in V(A)$.

5.3 Lemma. *If the class $\mathcal{S}$ satisfies the separation axiom (S4), then (Z6) holds for $V'(A)$.*

Proof. The proof is the same as the proof of decidable separation (Z6') in Proposition 5.1, except that the mono $\{z \in V'(A) \mid \varphi(z)\} \rightarrowtail V'(A)$ is now shown to be small by using (S4) (rather than the earlier assumption that this mono has a complement).

5.4 Lemma. *If the natural numbers object of $\mathcal{C}$ is small (as in (S5)), then (Z8) holds for $V'(A)$.*

Proof. There is a unique map $f : \mathbb{N} \to V(A)$ in $\mathcal{C}$ for which $f(0) = \emptyset$ and $f(n+1) = f(n) \vee s(f(n))$. When $\mathbb{N}$ is small, one can define an element $x \in V(A)$ by

$$x = \bigvee_{n \in \mathbb{N}} f(n) \; : \; 1 \to V(A).$$

This x witnesses the validity of the axiom of infinity.

Putting the results 5.1–4 together, we obtain the following two theorems.

5.5 Theorem. *Let $\mathcal{C}$ be a Boolean pretopos with a natural numbers object, equipped with a class of small maps $\mathcal{S}$ satisfying (S3) and (S5). Then for any object A in $\mathcal{C}$, with associated free Zermelo-Fraenkel algebra $V(A)$, the object $V'(A) = A + V(A)$ is a model of ZFA. In particular, the initial algebra V is a model of ZF.*

And in the non-Boolean case:

5.6 Theorem. *Let $\mathcal{C}$ be a Heyting pretopos with a natural numbers object, equipped with a class $\mathcal{S}$ of small maps satisfying (S3–5). Then $V'(A)$ is a model of IZFA. In particular, the free algebra V is a model of IZF.*

Chapter III

Existence Theorems

Let $\mathcal{C}$ be a suitable pretopos equipped with a class of small maps $\mathcal{S}$, as in Chapter II. In this chapter, we will make the additional assumption that the pretopos $\mathcal{C}$ has a subobject classifier (cf. Appendix B). This will be seen to imply that the various kinds of free Zermelo-Fraenkel algebras considered in the previous chapter exist in $\mathcal{C}$. Using the theory of bisimulations for the category of forests, we will give explicit constructions of the free ZF-algebra V, and of the different algebras of ordinals, O, N and T, discussed in Chapter II.

§1   Open maps and (bi-)simulations

Let $\mathcal{E}$ be a suitable category equipped with a distinguished class of open maps $\mathcal{O}$. (Thus $\mathcal{O}$ is a class of arrows in $\mathcal{E}$ satisfying the axioms (A1–6) from Section I.1.) In the examples to be developed, $\mathcal{E}$ will be a category of presheaves , equipped with its canonical class of open maps: recall from Joyal-Moerdijk(1994) that for presheaves on a small category I, a map $\varphi : P \to Q$ is said to be open iff for any arrow $u : j \to i$ in I, the square on the left is a quasi-pullback:

$$
\begin{array}{ccc}
P(i) \xrightarrow{\varphi_i} Q(i) & \qquad & z \dashrightarrow x \\
\downarrow{\scriptstyle P(u)} \qquad \downarrow{\scriptstyle Q(u)} & & \vert \qquad \downarrow \\
P(j) \xrightarrow[\varphi_j]{} Q(j) & & y \longrightarrow \varphi(y).
\end{array}
\tag{1}
$$

This means that for any $x \in Q(i)$ and any $y \in P(j)$ with $\varphi(y) = Q(u)(x)$ there exists a point $z \in P(i)$ such that $P(u)(z) = y$ and $\varphi(z) = x$, as on the right. Note that this canonical class of open maps is closed under infinite sums; i.e., infinite versions of axioms (A4) and (A5) hold.

A diagram (span)

$$\text{(2)}$$

in the category $\mathcal{E}$ is said to be a *simulation from E to F* if α is an open surjection while β is an open map. In this case we also write

$$S : E \leq F.$$

If α is an open surjection but β is just any map in the category $\mathcal{E}$ then S is said to be a *weak simulation* , denoted

$$S : E \leq_w F.$$

A *bisimulation* between E and F is a span (2) in which *both* α and β are open surjections. In this case we write

$$S : E \sim F.$$

Many different examples of this abstract notion of bisimulation are discussed in Joyal-Nielsen-Winskel(1993).

Observe that if $S : E \leq F$ and $T : F \leq E$ are simulations, one can construct an obvious bisimulation $S + T^{op} : E \sim F$. In other words,

$$S : E \leq F \text{ and } T : F \leq E \Rightarrow S + T^{op} : E \sim F. \qquad (3)$$

Also notice that (by the quotient axiom (A6) for open maps), for any (bi-)simulation of the form (2), the image $S' = (\alpha, \beta)(S) \subseteq E \times F$ is again a (bi-)simulation from E of F. Thus, if we are interested in the mere *existence* of (bi-)simulations between E and F, we may restrict our attention to subobjects S of $E \times F$.

There is an obvious "unit" bisimulation $E : E \sim E$, given by the identity span

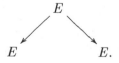

Furthermore, the composition of spans by pullback,

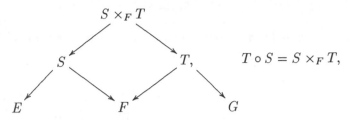

$$T \circ S = S \times_F T,$$

defines a composition on (weak/bi-)simulations, having the following evident properties:

1.1 Lemma. (i) *If $S : E \le F$ and $T : F \le G$ then $T \circ S : E \le G$.*
(ii) *If $S : E \sim F$ and $T : F \sim G$ then $T \circ S : E \sim G$.*
(iii) *If $S : E \le_w F$ and $T : F \le_w G$ then $T \circ S : E \le_w G$.*

Proof. Clear from the fact that open maps are stable under pullback and composition (axioms (A1) and (A2) in Chapter I, §1).

There is a preorder on the objects of $\mathcal{E}$, defined by

$$E \le F \quad \text{iff} \quad \exists S \subseteq E \times F \ \text{such that} \ S : E \le F. \tag{4}$$

The associated equivalence relation $E \sim F$ iff $E \le F$ and $F \le E$ is given by bisimulation (cf. (3)):

$$E \sim F \quad \text{iff} \quad \exists S \subseteq E \times F \ \text{such that} \ S : E \sim F. \tag{5}$$

We will denote by $[E]$ the equivalence class of an object E of $\mathcal{E}$.

There is also a larger preorder, defined from weak simulation in the same way:

$$E \le_w F \quad \text{iff} \quad \exists S \subseteq E \times F \ \text{such that} \ S : E \le_w F. \tag{6}$$

The associated equivalence relation is denoted by $E \sim_w F$, and we write $[E]_w$ for the equivalence class of an object E of $\mathcal{E}$.

In this way, we obtain two partial orders $\mathcal{E}/\sim$ and $\mathcal{E}/\sim_w$ from the category $\mathcal{E}$ and the class of open maps O. We next observe that sums in $\mathcal{E}$ give suprema in these partial orders.

1.2 Lemma. *For any coproduct $F = \sum F_i$ in the category $\mathcal{E}$ and any object G of $\mathcal{E}$:*
(i)   *Each coproduct inclusion $F_i \hookrightarrow F$ gives a simulation $F_i : F_i \le F$.*

(ii) *If $S_i : F_i \leq G$ are simulations, then so is $\sum S_i : \sum F_i \leq G$.*
(iii) *If $S_i : F_i \leq_w G$ then $\sum S_i : \sum F_i \leq_w G$.*

Proof. From the axioms for open maps (with infinite versions of (A4) and (A5)) it is clear that

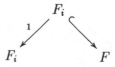

is a simulation, and that a family of (weak) simulations as below on the left is transformed into a (weak) simulation as on the right:

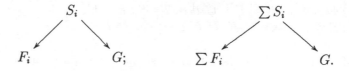

  1.3 Corollary. *For $F = \sum F_i$ and G as above,*
(i) $[F] \leq [G]$ *iff* $[F_i] \leq [G]$ *for each i,*
(ii) $[F]_w \leq_w [G]_w$ *iff* $[F_i]_w \leq_w [G]_w$ *for each i.*
In other words,

$$\bigvee_i [F_i] \;=\; [\textstyle\sum F_i] \quad in \;\; \mathcal{E}/\sim,$$

$$\bigvee_i [F_i]_w \;=\; [\textstyle\sum F_i]_w \quad in \;\; \mathcal{E}/\sim_w.$$

  Let $\Phi : \mathcal{E} \to \mathcal{E}$ be any endomorphism which preserves *open surjections*. Then Φ induces a well-defined operation on equivalence classes:

$$\Phi : \mathcal{E}/\sim \;\to\; \mathcal{E}/\sim . \tag{7}$$

Thus $\mathcal{E}/\sim$ becomes a Zermelo-Fraenkel algebra with successor Φ. Similarly, Φ induces a *monotone* operation

$$\Phi : \mathcal{E}/\sim_w \;\to\; \mathcal{E}/\sim_w, \tag{8}$$

thus making $\mathcal{E}/\sim_w$ into a Zermelo-Fraenkel algebra with monotone successor. In this chapter, we will show how the free ZF-algebras considered in Chapter II can be obtained by a suitable internal version of this construction in the ambient pretopos $\mathcal{C}$ with its class of small maps $\mathcal{S}$.

§2 Forests

We work in the ambient category $\mathcal{C}$ equipped with a class of small maps, as in earlier chapters. Recall that in this chapter, we assume that $\mathcal{C}$ has a subobject classifier Ω. This implies in particular that for every small object S the power-object $\mathcal{P}(S) = \Omega^S$ exists, since small objects are exponentiable. When arguing about structures in the category $\mathcal{C}$, we shall often exploit the first order logic of $\mathcal{C}$, and write informally, as if $\mathcal{C}$ were the category of sets.

Let N be the natural numbers object of $\mathcal{C}$. One can view N as a category, with a unique arrow $n \to m$ iff $n \leq m$, as usual. A *forest* is by definition an (internal) presheaf on N. Such a presheaf F can be represented informally as a sequence of objects and "restriction" maps in $\mathcal{C}$,

$$F = (F_0 \xleftarrow{r_0} F_1 \xleftarrow{r_1} F_2 \leftarrow \cdots) .$$

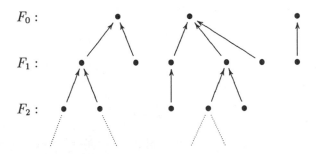

For two such forests F and G, a *map* between forests $\varphi : F \to G$ is a natural transformation; i.e., φ is a sequence of functions $\varphi_n : F_n \to G_n$ which commute with the restriction functions of F and G. This defines a category

$$(\text{Forests})_{\mathcal{C}}$$

of forests and maps between them in $\mathcal{C}$. We note that this category is a Grothendieck topos, relative to $\mathcal{C}$. In particular, it has $\mathcal{C}$-indexed sums, which are constructed "pointwise".

A forest is said to be (internally) *connected* if it cannot be written as a sum $F = \sum_{i \in I} F_i$ (for $I \in \mathcal{C}$) except in the trivial way $F = F + 0$. A connected forest is called a *tree* . Thus F is a tree iff F_0 is "the" terminal object of $\mathcal{C}$; the unique element of F_0 will then be called the *root* of the tree. Every forest is (isomorphic to) a sum of trees, its connected components. The trees form a full subcategory $(\text{Trees})_{\mathcal{C}}$ of the category of forests. Note that these categories (Trees) and (Forests) are equivalent categories, by the operations

$$(\text{Forests}) \underset{C}{\overset{S}{\rightleftarrows}} (\text{Trees}) : \tag{1}$$

Here C chops off the root of the tree,

$$C(F)_n = F_{n+1},$$

and the *successor operation* S adds a new root to a forest,

$$S(F)_0 = \{r\} \cong 1 \ , \ \ S(F)_{n+1} = F_n.$$

2.1 Notation. For a forest $F = (F_0 \overset{r_0}{\leftarrow} F_1 \leftarrow \cdots)$ we will also write F for the "total space" $\sum_{n \geq 0} F_n$. For $x \in F_n$ and $y \in F_m$ we write $y \leq x$ if $n \leq m$ and $r_n \circ \cdots \circ r_{m-1}(y) = x$; if furthermore $n < m$ we write $y < x$, and if $n + 1 = m$ we write $y \lessdot x$. In this case y is said to be an immediate predecessor of x. (Thus, the trees grow downwards from the root.) For each $x \in F_n$ there is a tree $\downarrow(x)$ defined by

$$\downarrow(x)_k = \{y \in F_{n+k} \mid y \leq x\},$$

with the obvious restriction maps $\downarrow(x)_{k+1} \to \downarrow(x)_k$ inherited from F. Note that there is a map of forests

$$i_x : \downarrow(x) \to F,$$

defined for $y \in \downarrow(x)_k$ by $i_x(y) =$ the unique $z \in F_k$ so that $y \leq z$.

As a presheaf category, the category of forests in $\mathcal{C}$ comes equipped with a canonical class of open maps. Explicitly, a map $\varphi : F \to G$ is open iff (it is internally valid in $\mathcal{C}$ that) for any $x \in F_n$ and $y \in G_{n+1}$ with $y \lessdot \varphi(x)$ there exists a $z \in F_{n+1}$ with $z \lessdot x$ and $\varphi(z) = y$.

The following lemma for the functor S, defined in (1) above, is clear:

2.2 Lemma. *The successor operation S preserves open surjections.*

Thus the functor S passes to a well-defined operation on (weak) bisimulation classes, defined as in the previous section by $S([F]) = [S(F)]$ (respectively, $S([F]_w) = [S(F)_w]$).

We will now restrict our attention to small and well-founded forests.

2.3 Definition. A forest $F = (F_0 \xleftarrow{r_0} F_1 \leftarrow \cdots)$ is said to be *small* if its total space $F = \sum_{n \geq 0} F_n$ is a small object (i.e. $F \to 1$ belongs to $\mathcal{S}$) and each restriction map r_i is small. (Or in more global terms, the restriction map $r : s^*(F) \to F$ is small, where $s : \mathsf{N} \to \mathsf{N}$ is the successor of the natural numbers object and s^* denotes pullback along s.)

Note that this implies that F_0, as a complemented subobject of F, is small. The fact that the restriction maps r_i are small can be rephrased by saying that for any $x \in F$, the set $\{y \in F \mid y \lessdot x\}$ is small.

For a small forest F, the power-set $\Omega^F = \mathcal{P}(F)$ exists in $\mathcal{C}$, since F is exponentiable. Thus the notion of a small *well-founded* forest in $\mathcal{C}$ can be defined, using the first order logic of $\mathcal{C}$, in the usual way:

2.4 Definition.(i) A subobject $A \subseteq F$ is said to be *inductive* if

$$\forall x \in F[\forall y \in F(y \lessdot x \Rightarrow y \in A) \Rightarrow x \in A].$$

(ii) A small forest F is said to be *well-founded* if

$$\forall A \in \mathcal{P}(F)[A \text{ is inductive} \Rightarrow A = F].$$

2.5 Lemma. *Let $\varphi : F \to G$ be a map between small forests.*
(i) *If G is well-founded then so is F.*
(ii) *If φ is an open surjection and F is well-founded then so is G.*

Proof. (i) Let $A \subseteq F$ be inductive. Then $\forall_\varphi(A) = \{x \in G \mid \varphi^{-1}(x) \subseteq A\}$ is inductive in G, so $\forall_\varphi(A) = G$ since G is assumed well-founded. But then $A = F$, showing F well-founded.

(ii) Let $B \subseteq G$ be inductive. Then $\varphi^{-1}(B) \subseteq F$ is also inductive, precisely because φ is open. Thus $\varphi^{-1}(B) = F$ since F is assumed well-founded. Hence $B = G$ since φ is surjective. This shows G well-founded.

2.6 Lemma. (i) *If F_i ($i \in I$) is a collection of well-founded small forests indexed by a small object $I \in \mathcal{C}$, then $\sum_{i \in I} F_i$ is again small and well-founded.*
(ii) *If F is a well-founded small forest then so is its successor $S(F)$.*
(iii) *If F is a well-founded small forest, then for any point $x \in F$ the tree $\downarrow(x)$ is again small and well-founded.*

Proof. By the definability of small maps (Chapter I, Proposition 1.6), we can argue "internally" about smallness, using the first order logic of $\mathcal{C}$. Then (i) and (ii) are straightforward. For (iii), let $B = \{x \in F \mid \downarrow(x) \text{ is}$

small and well-founded}. It suffices to show that B is inductive. But if $y \in B$ for every $y \lessdot x$, then also $x \in B$ since

$$\downarrow(x) \cong S(\sum_{y \lessdot x} \downarrow(y))$$

and we can apply parts (i) and (ii) of the lemma.

§3   Height functions

Let (L, σ) be a Zermelo-Fraenkel algebra with successor σ. An *L-valued height function* on a small forest F is a map $h : F \to L$ such that

$$h(x) = \sigma(\bigvee_{y \lessdot x} h(y)) \qquad (1)$$

for any $x \in F$. Notice that since F is assumed small, the sup occurring in this equation is over a small set, hence exists in L.

3.1 Lemma. *For any well-founded small forest F, and any ZF-algebra (L, σ), there exists a unique L-valued height function $h_F : F \to L$.*

Proof. For uniqueness, let $h, k : F \to L$ be two height functions, and let $U = \{x \in F \mid h(x) = k(x)\}$. From the identity (1) for h and for k it is clear that U is inductive. Since F is well-founded, $U = F$, hence $h = k$.

For existence, let $E = \{x \in F \mid$ there exists a height function $\downarrow(x) \to L\}$. If $x \in E$ then this height function $\downarrow(x) \to L$ is unique by the first part of the proof, since $\downarrow(x)$ is small and well-founded by Lemma 2.6 (iii). So we can denote it by $h_x : \downarrow(x) \to L$. To complete the proof, it suffices to show that E is inductive. But if $x \in F$ is such that $y \in E$ whenever $y \lessdot x$, then we can define $h_x : \downarrow(x) \to L$ by $h_x(z) = h_y(z)$ if $z \leq y \lessdot x$, and $h_x(x) = \sigma(\bigvee_{y \lessdot x} h_y(y))$. This proves the lemma.

3.2 Definition. Let F be a well-founded small forest, and L a ZF-algebra. If $h_F : F \to L$ denotes the unique height function, the *L-valued height* of F is defined by

$$L\text{-}height\,(F) := \bigvee_{x \in F_0} h_F(x).$$

Observe that this L-height satisfies the following obvious identities with respect to small sums and successor:

$$L\text{-}height(\Sigma F_i) = \bigvee_i (L\text{-}height(F_i)), \qquad (2)$$

$$L\text{-}height(S(F_i)) \;=\; \sigma(L\text{-}height(F)). \tag{3}$$

These identities determine the height:

3.3 Proposition. *Let H be any operation assigning to each well-founded small forest F an element $H(F) \in L$, such that H is invariant under isomorphism between forests, transforms small sums of forests into sups (as in (2)) and such that H preserves the successor (as in (3)). Then for any small well-founded forest F one has $H(F) = L\text{-}height(F)$.*

Proof. For a small well-founded forest F, with unique L-valued height function $h_F : F \to L$, let $A = \{x \in F \mid H(\downarrow(x)) = h_F(x)\}$. Then the isomorphism $\downarrow(x) \cong S(\sum_{y<x}\downarrow(y))$ shows that A is inductive. Thus $A = F$. But then

$$H(F) = H(\sum_{x \in F_0} \downarrow(x)) \;=\; \bigvee_{x \in F_0} h_F(x) = L\text{-}height(F).$$

For height functions into a Zermelo-Fraenkel algebra (L, σ) with a monotone successor $(x \le y \;\Rightarrow\; \sigma(x) \le \sigma(y))$, we observe the following special properties:

3.4 Lemma. *Let F be a well-founded small forest, and let $h_F : F \to L$ be its unique height function into a ZF-algebra (L, σ). If σ is monotone, then for any $x, y \in F$*
(i) $h_F(x) \le \sigma h_F(x)$,
(ii) $x \le y \;\Rightarrow\; h_F(x) \le h_F(y)$.

Proof. (i) Let $A = \{x \in F \mid h_F(x) \le \sigma h_F(x)\}$. It suffices to show that A is inductive. Suppose $h_F(y) \le \sigma h_F(y)$ for all $y \lessdot x$. Then

$$\bigvee_{y \lessdot x} h_F(y) \;\le\; \bigvee_{y \lessdot x} \sigma h_F(y)$$

$$\le\; \sigma(\bigvee_{y \lessdot x} h_F(y)),$$

the latter since σ is monotone. Applying σ to this inequality gives

$$\sigma(\bigvee_{y \lessdot x} h_F(y)) \;\le\; \sigma^2(\bigvee_{y \lessdot x} h_F(y)),$$

or $h_F(x) \le \sigma \, h_F(x)$. This shows A is inductive.
(ii) It suffices to consider the case where $y \lessdot x$. (Then the general case for $y \le x$ with $y \in F_m$ and $x \in F_n$ follows by induction on $m - n$.) But for

$y \lessdot x$, part (i) gives

$$h_F(y) \leq \sigma \, h_F(y) \leq \sigma(\bigvee_{p \lessdot x} h_F(p)) = h_F(x).$$

This proves the lemma.

Next, we consider invariance properties of height functions under (weak) simulations. Below, (L, σ) denotes a fixed ZF-algebra, and $h_F : F \to L$ denotes the unique height function, for any small well-founded forest F.

3.5 Proposition. *Let $F \xleftarrow{\alpha} S \xrightarrow{\beta} G$ be a span in the category of forests, and assume F and G are small and well-founded.*
(i)   *If S is a simulation from F to G then*

$$\forall s \in S : h_F(\alpha(s)) = h_G(\beta(s)).$$

(ii) *If S is a weak simulation from F to G, and if the successor σ in the ZF-algebra is monotone, then*

$$\forall s \in S : h_F(\alpha(s)) \leq h_G(\beta(s)).$$

Proof. (i) We show that for any $x \in F$

$$\forall s \in S \ (\alpha(s) = x \ \Rightarrow \ h_F(x) = h_G(\beta(s))) \tag{4}$$

by induction on the well-founded tree F. Suppose (4) holds for all $y \in F$ with $y \lessdot x$. To show that (4) holds for x, take any $s \in S$ with $\alpha(s) = x$. Since α is an open map, there exists for any $y \lessdot x$ a $t \in S$ with $t \lessdot s$ and $\alpha(t) = y$. Then $\beta(t) \lessdot \beta(s)$, and $h_F(y) = h_G(\beta(t))$ by the induction hypothesis. This shows that

$$\forall y \lessdot x \ \exists z \lessdot \beta(s) \ : \ h_F(y) = h_G(z).$$

A symmetric argument, using openness of the map β, similarly shows that

$$\forall z \lessdot \beta(s) \ \exists y \lessdot x \ : \ h_F(y) = h_G(z).$$

It follows that

$$\bigvee_{y \lessdot x} h_F(y) \ = \ \bigvee_{z \lessdot \beta(s)} h_G(z),$$

and hence, by applying σ to this identity, that

$$h_F(x) \ = \ h_G(\beta(s)).$$

This shows that (4) holds for x, as required.

(ii) We show that for any $x \in F$,

$$\forall s \in S(\alpha(s) = x \Rightarrow h_F(x) \leq h_G(\beta(s))), \tag{5}$$

again by induction on $x \in F$. If (5) holds for all $y \lessdot x$, then, using that α is open, we can show exactly as in part (i) that for any $s \in S$ with $\alpha(s) = x$

$$\forall y \lessdot x \;\exists z \lessdot \beta(s) \;:\; h_F(y) \leq h_G(z).$$

Thus

$$\bigvee_{y \lessdot x} h_F(y) \leq \bigvee_{z \lessdot \beta(s)} h_G(z).$$

Applying the monotone operator σ yields

$$h_F(x) \leq h_G(\beta(s)),$$

so that (5) holds for x.

3.6 Corollary. *Let F and G be well-founded small trees.*

(i) *If there exists a simulation $S : F \leq G$ then $L\text{-height}\,(F) \leq L\text{-height}(G)$.*

(ii) *If there exists a bisimulation $S : F \sim G$ then $L\text{-height}\,(F) = L\text{-height}(G)$.*

(iii) *Assume L has a monotone successor σ. If there exists a weak simulation $S : F \leq_w G$ then $L\text{-height}(F) \leq L\text{-height}(G)$.*

Proof. (i) Write $\alpha : S \twoheadrightarrow F$ and $\beta : S \to G$ for the maps in the simulation. Since α is surjective, $\forall x \in F_0 \,\exists s \in S \,\alpha(s) = x$. Hence by 3.5(i), with $y = \beta(s)$,

$$\forall x \in F_0 \;\exists y \in G_0 \; h_F(x) = h_G(y).$$

From this the inequality $L\text{-height}(F) \leq L\text{-height}(G)$ immediately follows, by Definition 3.2.

Part (ii) follows from (i). The proof of (iii) is the same as that of (i), now using 3.5(ii).

§4 Construction of V and O

In this section we will prove the existence of the initial ZF-algebra V considered in Chapter II, Section 1. In fact we will give an explicit construction of V from the universal small forest. This construction is a straightforward consequence of the properties of small forests and height functions considered before. Analogous constructions will also prove the existence of the algebra O

of ordinals, which is free on a monotone successor, and of the algebra $V(A)$, which is free on an object A of generators. We recall from the beginning of this chapter that these existence proofs assume that the ambient pretopos $\mathcal{C}$ has a subobject classifier Ω, and hence a power-object $\mathcal{P}(S)$ for each small object S.

4.1 Lemma. *There exists a universal small well-founded forest , denoted $p : F \to U$ (with forest structure $\varepsilon : F \to \mathsf{N}$ and $\alpha : s^*(F) \to F$).*

Proof. We will show that this is a consequence of Chapter I, Corollary 2.5. For the natural numbers object N, denote by $G(\mathsf{N})$ the graph with $G(\mathsf{N})_0 = \mathsf{N}$ as space of vertices, and an edge $n \to m$ iff $m = n + 1$. Then N as a category is free on the graph $G(\mathsf{N})$, so a presheaf on N is the same thing as a $G(\mathsf{N})$-object. Since the codomain map of $G(\mathsf{N})$ is the successor map $s : \mathsf{N} \to \mathsf{N}$, which – being a monomorphism with a complement – is a small map, Corollary I.2.5 gives that there exists a universal small $G(\mathsf{N})$-object with small action map, call it $D = (D \to W, \varepsilon, \alpha)$. Then D is the universal small forest. Now by exponentiability of small maps and the existence of a subobject classifier Ω, well-foundedness for small forests is definable in the first order logic of the pretopos $\mathcal{C}$ (cf. Definition 2.4). Thus there is a subobject $U \subseteq W$ in $\mathcal{C}$, defined as

$$U = \{x \in W \mid D_x \text{ is well-founded}\}.$$

Let $F \to U$ be the pullback of the universal small forest $D \to W$ along this inclusion $U \hookrightarrow W$. Then $F \to U$ is the desired universal small well-founded forest.

As usual, we shall denote the fiber of this universal small well-founded forest $p : F \to U$ over a point $x \in U$ by $F_x = p^{-1}(x)$. Thus, informally, universality means that each F_x is a small well-founded forest, and that every other such forest G is isomorphic to F_x for some point $x \in U$.

Next, we define a subobject $B \subseteq U \times U$ by

$$B = \{(x,y) \mid \exists S \in \mathcal{P}(F_x \times F_y) : S \text{ is a bisimulation between } F_x \text{ and } F_y\}.$$

Notice that B is a well-defined object of $\mathcal{C}$, since $F_x \times F_y$ is small so that $\mathcal{P}(F_x \times F_y)$ exists. Also, by the properties of bisimulations discussed in §1, B is an equivalence relation on U.

4.2 Lemma. *The quotient U/B has the structure of a Zermelo-Fraenkel* *algebra* $(U/B, \bigvee, s)$.

Proof. As before, we will argue in this proof using the internal logic of the category $\mathcal{C}$ in an informal way, as if $\mathcal{C}$ were the category of sets. In the proof,

$$q : U \to U/B$$

will denote the quotient map.

To begin with, define another subobject $P \subseteq U \times U$, by

$$P = \{(x, y) \mid \exists S \in \mathcal{P}(F_x \times F_y) : S \text{ is a simulation from } F_x \text{ to } F_y\}.$$

Then, by the properties of (bi-)simulations discussed in Section 1, P is a preorder on U and B is the associated equivalence relation, $B = \{(x, y) \in U \times U \mid (x, y) \in P \text{ and } (y, x) \in P\}$. Thus P induces a partial order $\leq$ on the quotient U/B.

Next, we show that for this partial order all small suprema exist in U/B. To this end, let $A \subseteq U/B$ be any small subset. By the Collection Axiom (A7) for small maps, there exists a small subset $A' \subseteq U$ with $q(A') = A$. Let $G = \Sigma_{x \in A'} F_x$. Then G is a small well-founded forest. So by universality of $F \to U$, there exists a point $y \in U$ for which there is an isomorphism $G \cong F_y$. Define $\bigvee A = q(y) \in U/B$. It follows from the properties of simulation in Lemma 1.2 and Corollary 1.3 that this point $q(y)$ is indeed the supremum of A in $(U/B, \leq)$.

Finally, we define a successor operation $s : U/B \to U/B$. For this, pick any point $\xi \in U/B$, and let $x \in U$ be any point such that $q(x) = \xi$. The successor of the corresponding small well-founded forest F_x gives another forest $S(F_x)$, which is again small and well-founded (cf. Lemma 2.6 (ii)). By universality of $F \to U$, there is thus a point $y \in U$ with $S(F_x) \cong F_y$. Define $s(\xi) = q(y)$ for this y. To see that this gives a well-defined map $s : U/B \to U/B$, note that if x' is any other point with $q(x') = \xi$, then there exists a bisimulation $F_x \sim F_{x'}$. Hence, by Lemma 2.2, also $S(F_x) \sim S(F_{x'})$. Thus if $y, y' \in U$ are such that $S(F_x) \cong F_y$ and $S(F_{x'}) \cong F_{y'}$, there exists a bisimulation $F_y \sim F_{y'}$. But then $q(y) = q(y')$. This shows that q is well-defined, and completes the description of the ZF-algebra structure on U/B.

4.3 Theorem. $(U/B, \bigvee, s)$ *is the initial ZF-algebra in* $\mathcal{C}$.

Proof. Let (L, σ) be another ZF-algebra. We define a homomorphism $\varphi : U/B \to L$ as follows. For $\xi \in U/B$, let $x \in U$ be any point with $q(x) = \xi$, and let F_x be the corresponding small well-founded forest. Define

$$\varphi(\xi) := L\text{-}height(F_x) \in L.$$

Note that this is independent of the choice of x. For if $q(x) = \xi = q(x')$, then there exists a bisimulation $F_x \sim F_{x'}$; hence, by Corollary 3.6 (ii), F_x and $F_{x'}$ have the same height.

Using the construction of suprema and successors in U/B from the proof of Lemma 4.2, we check that φ is a homomorphism of ZF-algebras. First, if $\xi \leq \zeta$ in U/B and $\xi = q(x)$ while $\zeta = q(y)$, then $(x,y) \in P$; i.e. there exists a simulation from F_x to F_y. Thus by Corollary 3.6 (i) we have $L\text{-}height(F_x) \leq L\text{-}height(F_y)$, or $\varphi(\xi) \leq \varphi(\zeta)$. This shows that φ is monotone. Next, since suprema in U/B are defined by sums of small well-founded forests, it follows from the identity (2) in §3 that φ preserves small suprema. Similarly, since the successor $s : U/B \to U/B$ is defined by the usual successor of small well-founded forests, identity (3) in §3 shows that φ preserves this successor.

Finally, for uniqueness, suppose $\psi : U/B \to L$ is any other homomorphism. Let G be any small well-founded forest. By universality of $F \to U$, there exists a point $x \in U$ for which $F_x \cong G$. We thus obtain an element $H(F) := \psi(q(x)) \in L$. Note that $H(G)$ does not depend on the choice of x; indeed, if $F_x \cong G \cong F_{x'}$ then $F_x \sim F_{x'}$, hence $q(x) = q(x')$. Furthermore, by the explicit description of sups and the successor on U/B, it follows that $H(\Sigma G_i) = \bigvee H(G_i)$ for any small family $\{G_i\}$ of small well-founded forests G_i, while $H(S(G)) = \sigma(H(G))$ for any small well-founded forest G. By uniqueness (Proposition 3.3), we must have $H(G) = L\text{-}height(G)$. Thus $\varphi = \psi$.

This completes the proof of the theorem.

This theorem provides a construction of the cumulative hierarchy V. In a completely analogous way, one can also construct the algebra O of ordinal numbers, free on a monotone successor. Define a subobject $P_w \subseteq U \times U$ by

$$P_w = \{(x,y) \mid \exists S \in \mathcal{P}(F_x \times F_y) : S \text{ is a weak simulation from } F_x \text{ to } F_y\}.$$

By the properties of weak simulations derived in §1, P_w is a preorder on U. Let

$$B_w = \{(x,y) \mid (x,y) \in P_w \text{ and } (y,x) \in P_w\}$$

be the associated equivalence relation. Exactly as in the proof of Lemma 4.2, sums and successor of small well-founded forests define on U/B_w the structure of a ZF-algebra. Observe that this successor on U/B_w is monotone, because the successor of forests preserves weak simulations (cf. Lemma 2.2).

4.4 Theorem. *The ZF-algebra U/B_w is the initial ZF-algebra with a* *monotone successor.*

Proof. The proof is completely analogous to that of Theorem 4.3, now using the properties of height functions on forests with values in a ZF-algebra L with monotone successor (cf. Corollary 3.6(iii)).

4.5 Remark. In a similar way one can prove the existence of the free ZF-algebra $V(A)$ on an arbitrary object of generators $A \in \mathcal{C}$. For this, one uses small well-founded forests F equipped with a "labeling" of the nodes. Such a labeling consists of a complemented subset $E \subseteq F$, such that each $x \in E$ is an end-node of F (i.e. x has no immediate predecessors), together with a map $\lambda : E \to A$. There is a straightforward generalization of the notions of (bi-)simulation and height-functions (with values in a ZF-algebra L equipped with a map $A \to L$) to such labeled forests, and $V(A)$ can be constructed as the bisimulation quotient of the universal well-founded small forest equipped with a labeling in A. The details are analogous to those for the construction of V just presented.

§5   Construction of Tarski ordinals

Recall from Chapter II, §4, that the object of Tarski ordinals (T, r) is the initial poset with small weakly directed suprema and a monotone successor r. In this section we will show that (T, r) exists, by giving an explicit construction of T as a subobject of the algebra $O = U/B_w$ of ordinal numbers, constructed in the preceding section from weak simulation. Such a weak simulation between small well-founded forests is a diagram

$$(1)$$

where α is an open surjection and β is any map between forests. From these simulations, we defined a preorder $F \leq_w G$ on small well-founded forests. For a single such forest F, this induces a preorder $\leq_w$ on the points of F defined by

$$x \leq_w y \quad \text{iff} \quad \downarrow(x) \leq_w \downarrow(y).$$

A small well-founded forest F is said to be a *Tarski forest* if the following two conditions hold:

(i)   For each $x \in F$, the set $\{y \in F \mid y \lessdot x\}$ of immediate predecessors of x is weakly directed in the preorder $\leq_w$;

(ii) F_0 is weakly directed in this preorder.

Note the following obvious closure properties of the collection of Tarski forests:

5.1 Lemma. (i) *If F is a Tarski forest then so is its successor $S(F)$.*
(ii) *If $\{F_i\}$ is a small collection of Tarski forests, which is weakly directed for the preorder $\leq_w$, then ΣF_i is again a Tarski forest.*

Now consider a poset P with small weakly directed sups and a monotone successor σ. Define a *height function* from a Tarski forest F into such a poset P to be a map $h : F \to P$ such that for each $x \in F$, the set $\{h(y) \mid y \lessdot x\}$ is weakly directed, and moreover, such that

$$h(x) = \sigma(\bigvee\nolimits_{y \lessdot x} h(y)), \tag{2}$$

exactly as before. (This formula (2) makes sense since the sup occurring in it is weakly directed.) Evidently, such a height function is unique when it exists.

5.2 Lemma. *Let F and G be Tarski forests, and let $S : F \leq_w G$ be a weak simulation from F into G, as in (1). Suppose there exist height functions $h : F \to P$ and $k : G \to P$. Then*

$$\forall s \in S : h(\alpha(s)) \leq k(\beta(s)) \text{ in } P.$$

Proof. The proof of Proposition 3.5(ii) applies verbatim, since for Tarski forests F and G all the suprema occurring there are weakly directed.

5.3 Lemma. *For any Tarski forest F there exists a unique height function $h_F : F \to P$.*

Proof. As already noted, uniqueness is clear. For existence, let $E = \{x \in F \mid \exists\ P\text{-valued height function } h_x :\downarrow (x) \to P\}$. We show that E is inductive. Suppose $x \in F$ is such that $y \in E$ whenever $y \lessdot x$. Since F is a Tarski forest, Lemma 5.2 implies that the set $\{h_y(y) \mid y \lessdot x\}$ is weakly directed. Thus its supremum exists in P, and we can define $h_x :\downarrow (x) \to P$, by $h_x(z) = h_y(z)$ for $z \leq y \lessdot x$ and $h_x(x) = \sigma(\bigvee\{h_y \mid y \lessdot x\})$.

Now let $p : F \to U$ be the universal small well-founded forest, constructed in Lemma 4.1. Since the notion of Tarski forest can be expressed in the first order logic of the ambient category $\mathcal{C}$, we can define a subobject $U' \subseteq U$ by $U' = \{x \in U \mid F_x$ is a Tarski forest$\}$, and we can construct the universal Tarski forest by a pullback

$$
\begin{array}{ccc}
F' & \hookrightarrow & F \\
\downarrow & & \downarrow \\
U' & \hookrightarrow & U.
\end{array}
$$

Let B'_w be the equivalence relation on U', induced by the equivalence relation B_w on U considered in §4. Consider the quotient U'/B'_w. This quotient inherits a partial order $\leq_w$ from the preorder P_w on U, so that the inclusion $U' \hookrightarrow U$ induces an embedding of posets

$$
U'/B'_w \;\hookrightarrow\; U/B_w. \tag{3}
$$

It follows by Lemma 5.1, and the construction of suprema and successor on U/B_w in Section 4, that this successor restricts to an operation

$$
s' : \; U'/B'_w \to U'/B'_w,
$$

and that all weakly directed small sups exist in U'/B'_w and are preserved by this embedding (3).

5.4 Theorem. *The poset U'/B'_w, with its successor s' inherited from the successor $s : U/B_w \to U/B_w$, is the initial poset with weakly directed sups and a monotone successor.*

Thus $(U'/B'_w, s')$ is (isomorphic to) the object of Tarski ordinals.

Proof. The proof uses Lemma 5.3 and is otherwise completely analogous to the proof of Theorem 4.3.

§6 Simulation for Von Neumann ordinals

Theorems 3.1 and 3.4 of Chapter II provide constructions of the object $(N, \bar{s})$ of Von Neumann ordinals, as a retract of (V, s) and as the object of hereditarily transitive "sets" inside V. It is also possible to construct $(N, \bar{s})$ directly from well-founded small forests and (bi-)simulations. In this section, we present one such construction.

Besides the category of forests, we consider a larger category with as objects the sets (i.e., objects of $\mathcal{C}$) S equipped with a strict (irreflexive) partial order $<$, and with as arrows all functions which preserve this order. This category is denoted (Strict Orders)$_{\mathcal{C}}$. Note that this category has sums and pullbacks (computed in terms of the underlying sets), but no terminal object.

In this category of strict orders, there is a notion of *open map* analogous to the one for forests: A map $\varphi : (S, <) \to (T, <)$ between strict orders is said to be open if, for any $s \in S$ and $t \in T$ with $t < \varphi(s)$, there exists an $s' \in S$ with $s' < s$ and $\varphi(s') = t$.

There is an evident inclusion functor

$$I : (\text{Forests})_{\mathcal{C}} \quad \to \quad (\text{Strict Orders})_{\mathcal{C}},$$

which sends a forest F to its total space F with the order $<$ described in 2.1. This inclusion functor preserves open maps.

Now define an *order simulation* from a forest F to a forest G to be a diagram in this category of strict orders,

$$
\begin{array}{ccc}
 & T & \\
{}^{\alpha}\swarrow & & \searrow^{\beta} \\
I(F) & & I(G)
\end{array}
\qquad (1)
$$

where α and β are open maps and α is surjective. The diagram is said to be an *order bisimulation* if β is also surjective. For such a simulation or bisimulation, we use the notation

$$T : F \leq_o G \ , \quad \text{respectively } T : F \sim_o G.$$

These more general kinds of (bi-)simulations have exactly the same formal properties as the ones discussed in §1. In particular, order (bi-)simulations can be composed by pullback, and the sum of order simulations $T_i : F_i \leq_o G$ gives an order simulation $T = \sum T_i : \sum F_i \leq_o G$. There is also an evident successor operation S on the category of strict orders, defined by adding a new top element. The inclusion functor I commutes with this successor. Also note that S transforms an order bisimulation $I(F) \leftarrow B \to I(G)$ into an order bisimulation $I(SF) = S(IF) \leftarrow S(B) \to S(IG) = I(SG)$. Furthermore, we observe that there is for any forest F an obvious open embedding

$$I(F) \hookrightarrow I(SF),$$

such that

$$I(F) : F \leq_o S(F). \qquad (2)$$

It can now be shown that the Von Neumann ordinals can be constructed from the universal small well-founded forest $F \to U$ by factoring out the equivalence relation B_o on U given by

$$(x, y) \in B_o \text{ iff } \exists S \in F_x \times F_y : S \text{ is an order bisimulation between } F_x \text{ and } F_y.$$

Then U/B_o is partially ordered by the relation $\leq_o$ of order simulation, defined analogously to the partial order on U/B in §4. And exactly as before, small suprema in U/B_o exist, and are constructed from small sums of forests. Moreover, there is a well-defined successor operation $\bar{s} : U/B_o \to U/B_o$, induced by the successor operation $F \mapsto S(F)$ on small forests. But now note that, by (2), this successor is inflationary.

The proof that $(U/B_o, \bar{s})$ is the initial ZF-algebra with an inflationary successor is based on the following analog of Proposition 3.5(i). In this lemma, (L, σ) is a fixed ZF-algebra with inflationary successor ($x \leq \sigma(x)$), and for a small well-founded forest F the unique L-valued height function is denoted by $h_F : F \to L$.

6.1 Lemma. *Let F and G be well-founded small forests, and suppose that $I(F) \xleftarrow{\alpha} T \xrightarrow{\beta} I(G)$ is an order simulation from F to G. Then for all $t \in T$*

$$h_F(\alpha t) = h_G(\beta t).$$

Proof. First notice that (as in Lemma 3.4) h_F is order preserving. Indeed, if $y \lessdot x$ in F then

$$h_F(y) \leq \bigvee_{p \lessdot x} h_F(p)$$

$$\leq \sigma(\bigvee_{p \lessdot x} h_F(p))$$

$$= h_F(x).$$

It follows that $h_F(y) \leq h_F(x)$ whenever $y \leq x$ in F.

Next, we prove by induction on the well-founded small forest $F \times G$ that for all $(x, y) \in F \times G$:

$$\exists t \in T(x = \alpha(t) \text{ and } y = \beta(t)) \Rightarrow h_F(x) = h_G(y).$$

Suppose this is true for all $(p, q) \in F \times G$ with $p < x$ and $q < y$. To prove the assertion for (x, y), suppose $x = \alpha(t)$ and $y = \beta(t)$, for some $t \in T$. Consider any $p \in F$ with $p \lessdot x$. Since α is open, $p = \alpha(s)$ for some $s < t$ in T. Let $q = \beta(s)$. Then $q < y$, hence $q \leq r \lessdot y$ for some $r \in G$. Thus

$h_F(p) = h_G(q) \leq h_G(r)$. This shows that from mere openness of α it follows that

$$\forall p \lessdot x \; \exists \, r \lessdot y : h_F(p) \leq h_G(r),$$

and hence

$$\bigvee_{p \lessdot x} h_F(p) \; \leq \; \bigvee_{r \lessdot y} h_G(r).$$

A symmetric argument using openness of β yields the reverse inequality. Thus

$$\bigvee_{p \lessdot x} h_F(p) \; = \; \bigvee_{r \lessdot y} h_G(r).$$

Applying σ yields the desired identity $h_F(x) = h_G(y)$.

We conclude for an inflationary ZF-algebra (L, σ):

6.2 Corollary. *Let F and G be well-founded small forests. If $F \leq_o G$ then L-height$(F) \leq L$-height(G). Furthermore, if $F \sim_o G$ then L-height$(F) = L$-height(G).*

Exactly as before, one then derives the following result, analogous to Theorems 4.3, 4.4 and 5.4.

6.3 Theorem. *The quotient $(U/B_o, \bar{s})$ is the initial ZF-algebra with an inflationary successor.*

In other words, $(U/B_o; \bar{s})$ is isomorphic to the algebra $(N, \bar{s})$ of Von Neumann ordinals.

Chapter IV

Examples

In this short chapter we will briefly describe some of the simplest examples of classes of small maps. Before we embark on this, we should make two remarks: First, as yet not much is known about the various free ZF-algebras in most of these examples. We hope that future research will shed some light on their exact nature. Second, there are many other natural examples of classes of maps, not mentioned below, which satisfy some but not all of our axioms for a class of small maps. These examples are often of a geometric nature, and related to the examples of classes of open and étale maps considered in Joyal-Moerdijk(1994).

We begin with two obvious examples, related to sets and classes, and to Kuratowski finiteness, respectively.

§1   Sets and classes

The easiest examples are provided by (the usual models of) Zermelo-Fraenkel set theory. Let $\mathcal{C}$ be the category of sets (relative to a fixed model of set theory), and let κ be an infinite regular cardinal. Let $\mathcal{S}$ be the class of maps $f : Y \to X$ with the property that for each point $x \in X$ the fiber $f^{-1}(x)$ has cardinality strictly less than κ. This class $\mathcal{S}$ satisfies the axioms (A1–7) for open maps. It also satisfies the additional axioms (S1) and (S2) for small maps. To construct a universal small map $\pi : E \to U$, let S be a fixed set of cardinality κ, and let $U = \{A \subseteq S \mid \operatorname{card}(A) < \kappa\}$, while $E = \{(a, A) \mid a \in A \in U\}$. Define $\pi : E \to U$ to be the projection, $\pi(a, A) = A$. Observe that the axiom of choice is not needed for showing the universality of this map. Indeed, given a small map $f : Y \to X$, let $X' = \{(x, e, A) \mid x \in X, \ A \in U, \ e : A \xrightarrow{\sim} f^{-1}(x)\}$. Then the projection $X' \to X$ is surjective, and together with the projection $X' \to U$ it fits into a double pullback of the required form:

$$
\begin{array}{ccc}
Y & \xleftarrow{\;\varepsilon\;} Y' & \longrightarrow E \\
\downarrow & \downarrow & \downarrow{\scriptstyle \pi} \\
X & \longleftarrow X' & \longrightarrow U
\end{array}
$$

where $Y' = \{(x, e, A, a) \mid (x, e, A) \in X' \text{ and } a \in A\}$, and $\varepsilon(x, e, A, a) = e(a)$. The additional separation axiom (S4) also holds in this example, since $\mathcal{C}$ is Boolean.

As to the hypotheses for theorems II.5.5 and II.5.6, if $\kappa > \omega$ then the natural numbers object of $\mathcal{C}$ is small (S5), and if κ is inaccessible then $\mathcal{S}$ satisfies the power-set axiom (S3). As the initial algebra V one recovers the hierarchy V_κ, and all the algebras of ordinals coincide in this example.

A related example is obtained by letting $\mathcal{C}$ be the category of classes (say, in (a fixed model of) Gödel-Bernays set theory), and by defining a map $f : Y \to X$ to be small if each fiber $f^{-1}(x)$ is (isomorphic to) a set.

§2   Kuratowski finite maps

Let $\mathcal{E}$ be an elementary topos with a natural numbers object N. Call a map $f : Y \to X$ Kuratowski finite (K-finite) if f is Kuratowski finite as an object of $\mathcal{E}/X$ (cf. Johnstone(1977), p. 302). Let $\mathcal{K}$ be the class of these K-finite maps. This class $\mathcal{K}$ satisfies all the axioms for small maps. For the representability axiom (S2), define $\pi : E \to U$, using the internal logic of $\mathcal{E}$, as follows. Let

$$U = \{(n, R) \mid n \in \mathsf{N} , R \text{ an equivalence relation on } \{1, \cdots, n\}\}$$

(this object U can be constructed as a subobject of $\mathsf{N} \times \mathcal{P}\,(\mathsf{N} \times \mathsf{N})$),

$$E = \{(n, R, \xi) \mid (n, R) \in U , \xi \in \{1, \cdots, n\}/R\}$$

and let $\pi : E \to U$ be the projection.

In this example, a ZF-algebra is a semi-lattice (a poset L with a bottom element 0 and binary sups) with a successor. The initial such algebra V is the object of internally hereditarily finite sets.

Since every K-finite monomorphism is complemented, the algebra of ordinals O and the algebra of Von Neumann ordinals N coincide (cf. Chapter II, Remark 3.6), and are isomorphic to the natural numbers object N. Since every K-finite subset of N is weakly directed, the Tarski ordinals T are also given by the natural numbers object.

§3 Sheaves on a site

The construction of sets and classes in §1 can be generalized to the context of an arbitrary Grothendieck topos . This construction is related to forcing models of set theory, and their sheaf-theoretic construction discussed in Fourman (1980).

Consider the topos $Sh(\mathbf{C})$ of sheaves on a site $\mathbf{C}$, and assume that $\mathbf{C}$ has pullbacks, and that the topology of $\mathbf{C}$ is subcanonical. Let κ be a fixed infinite regular cardinal, with the property that every cover in the site $\mathbf{C}$ has cardinality strictly less than κ. (For example, if $Sh(\mathbf{C})$ is a coherent topos one can take $\kappa = \omega$.) Define a sheaf X on $\mathbf{C}$ to be κ-*small* if X is covered by a collection of representable sheaves of cardinality strictly less than κ, as in

$$\sum_{i \in I} C_i \twoheadrightarrow X \, , \quad C_i \in \mathbf{C}, \, \text{card}(I) < \kappa.$$

Here we have identified each object $C_i \in \mathbf{C}$ with the corresponding representable sheaf $\mathbf{C}(-, C_i)$. Next, define a map $f : Y \to X$ in the category $Sh(\mathbf{C})$ of sheaves to be κ-*small* if for any $C \in \mathbf{C}$ and any arrow $x : C \to X$, the fiber $f^{-1}(x)$, constructed as the pullback

$$\begin{array}{ccc} f^{-1}(x) & \longrightarrow & Y \\ \downarrow & & \downarrow {\scriptstyle f} \\ C & \xrightarrow{\;x\;} & X, \end{array}$$

is a κ-small sheaf.

3.1 Proposition. *The class of κ-small maps between sheaves satisfies the axioms (A1–7) and (S1), (S2) for small maps.*

For the proof of the axioms (A1-7) for open maps, one uses the following simple properties of this class.

3.2 Lemma. *Let $f : X \to C$ be any map into a representable sheaf.*
(i) *If f is κ-small, then so is each pullback $D \times_C X \to D$ along a map $D \to C$ in the site $\mathbf{C}$.*
(ii) *If $\{C_j \to C\}_j$ is a cover in the site $\mathbf{C}$ and each pullback $C_j \times_C X \to C_j$ is κ-small, then $f : X \to C$ is κ-small.*

Proof. (i) is clear by the assumption that $\mathbf{C}$ has pullbacks, and (ii) follows by our assumption that κ exceeds the cardinality of each cover in the site.

The proof of the axioms (A1–7), for open maps, for this class of κ-small maps is now straightforward. For the descent axiom (A3), use Lemma 3.2(ii). By way of example, we verify the collection axiom (A7). For this, consider a small map $f : Y \to X$ and an epi $p : Z \twoheadrightarrow Y$. Let $x : C \to X$ be any map from a representable object C. By assumption on f, there exists a quasi-pullback

$$
\begin{array}{ccc}
\sum_{i \in I} C_i & \xrightarrow{\ y\ } & Y \\
\downarrow & & \downarrow{\scriptstyle f} \\
C & \xrightarrow{\ \ x\ \ } & X
\end{array}
$$

for a set I of cardinality $< \kappa$. For each $i \in I$, consider for the composition $y_i : C_i \hookrightarrow \sum C_i \to Y$ the pullback $p^{-1}(y_i) = C_i \times_Y Z \twoheadrightarrow C_i$. Since this map is epi, there is a cover $\{\alpha_{i,j} : D_{i,j} \to C_i \mid j \in J_i\}$ with the property that each $\alpha_{i,j}$ factors through $p^{-1}(y_i) \to C_i$. The sum of all these covers gives a quasi-pullback diagram

$$
\begin{array}{ccccc}
D_x = \sum_{i,j} D_{i,j} & \longrightarrow & Z & \longrightarrow & Y \\
\downarrow & & & & \downarrow{\scriptstyle f} \\
C & \xrightarrow{\hspace{3cm} x \hspace{3cm}} & & & X,
\end{array}
$$

where the left-hand vertical arrow is small, by the assumption on the size of the covers in **C**. The sum of all these quasi-pullback squares, for all $x : C \to X$, gives a quasi-pullback square

$$
\begin{array}{ccccc}
\sum_x D_x & \longrightarrow & Z & \longrightarrow & Y \\
\downarrow & & & & \downarrow{\scriptstyle f} \\
\sum_x C & \xrightarrow{\hspace{3.5cm}} & & & X
\end{array}
$$

of the required form. This proves the validity of the collection axiom (A7).

The exponentiability axiom (S1) evidently holds; in fact, since $Sh(\mathbf{C})$ is a topos, every object in exponentiable.

For the representability axiom (S2), consider for each object $D \in \mathbf{C}$ the set $\mathcal{F}_D$ of all families $f = \{f_i : C_i \to D\}_{i \in I}$ of arrows into D where $\mathrm{card}(I) < \kappa$. For such a family f, we write $C_f = \sum_{i \in I} C_i$, and we also write $f : C_f \to D$ for the arrow induced by the f_i. Now let $\mathcal{R}_f$ be the set of fiberwise equivalence relations R on $C_f \to D$. (This means that R is an equivalence relation on

C_f so that $R \subseteq C_f \times_D C_f$.) Now define

$$U = \sum_{D \in \mathbf{C}} \sum_{f \in \mathcal{F}_D} \sum_{R \in \mathcal{R}_f} \mathbf{C}(-, D).$$

For each summation index D, f, R, let $E_{D,f,R} = C_f/R$ which is an object over D. Define E to be the sum of all these, which is then an object over U:

$$C_f \longrightarrow C_f/R \hookrightarrow E$$

$$D \hookrightarrow U \quad \text{(coproduct inclusion for } D, f, R).$$

To show that $E \to U$ is universal, consider any small map $Y \to X$. The codomain X can be covered by a family of representables $\alpha : D \to X$, and for each such D the pullback $D \times_X Y$ is a small sheaf. Thus $D \times_X Y$ is covered by a family of representables of cardinality $< \kappa$, say

$$\sum_{i \in I} C_i \overset{p}{\twoheadrightarrow} D \times_X Y.$$

Write $f_i : C_i \to D$ for the composition

$$C_i \hookrightarrow \sum C_i \overset{p}{\twoheadrightarrow} D \times_X Y \overset{\pi_1}{\to} D.$$

For the notation introduced above, we then have $C_f = \sum C_i$, and $f : C_f \to D$ is $\pi_1 \circ p$. Let $R = Ker(p)$ be the kernel pair of p. Then $R \in \mathcal{R}_f$ and $C_f/R \cong D \times_X Y$. All this fits into a diagram

$$Y \longleftarrow D \times_X Y \hookrightarrow E$$

$$X \longleftarrow D \hookrightarrow U$$

in which both squares are pullbacks; the map $D \hookrightarrow U$ here is the coproduct inclusion for D, f, R. Taking the sum, for all $\alpha : D \to X$ in the family of representables covering X, yields a double pullback diagram of the required form:

$$Y \longleftarrow \sum_\alpha D \times_X Y \longrightarrow E$$

$$X \longleftarrow \sum_\alpha D \longrightarrow U. \tag{2}$$

This proves the representability axiom for the class of κ-small maps in $Sh(\mathbf{C})$, and completes the proof of Proposition 3.1.

3.3 Remarks. (a) Note that in this example, an "infinite" version of axiom (A5) holds.

(b) For a coherent topos and $\kappa = \omega$, the κ-small objects (or maps) are precisely the quasi-coherent ones.

(c) The Separation Axiom (S4) holds for this class of κ-small maps when κ exceeds the cardinality of the set of all arrows in the site **C**.

(d) As to the other hypotheses for Chapter II, Theorem 5.5 and 5.6, if $\kappa > \omega$ and **C** has a terminal object then the natural numbers object in $Sh(\mathbf{C})$ is small (axiom (S5)); and if, in addition, κ is inaccessible, then the Power-set Axiom holds.

Finally, we note that there is an analogous construction of a class of small maps in the pretopos of sheaves of classes on **C**, where a map is small if all its fibers are sets.

§4 Realizability

This example is related to realizability models of set theory, cf. Friedman (1973), McCarty (1984). For our ambient category $\mathcal{C}$ we take the effective topos *Eff*, proposed by D. Scott and first described in Hyland (1982). Recall that there are adjoint functors

$$\Gamma : \textit{Eff} \;\; \rightleftarrows \;\; \textit{Sets} : \Delta, \tag{1}$$

with Γ left (sic!) adjoint to Δ, where Γ preserves finite limits (and, as a left adjoint, all colimits), and Δ preserves epimorphisms. Also recall from Robinson-Rosolini (1990) that the category *Eff* has enough projectives. In particular, if S is a separated object of *Eff*, then S has a canonical projective cover $\tilde{S} \twoheadrightarrow S$ with $\tilde{S} \subseteq S \times \mathbf{N}$ over S. Here $\mathbf{N}$ denotes the natural numbers object of *Eff*. The collection of projective objects in *Eff* is closed under finite products, and under double-negation-closed subobjects. Moreover, the functor Δ in (1) preserves projectives.

Now fix a regular cardinal $\kappa > \omega$, and define a map $Y \to X$ in *Eff* to be *κ-small* if there exists a diagram

$$\begin{array}{ccc} Q & \twoheadrightarrow & Y \\ \downarrow & & \downarrow \\ P & \twoheadrightarrow & X \end{array} \tag{2}$$

where $P \to X$ is an epimorphism, as indicated, and where $Q \to P \times_X Y$ is epi as well (i.e., the square is a quasi-pullback), and where moreover P and Q are both projective and $\Gamma Q \to \Gamma P$ is a κ-small map in *Sets*. (Recall from §1 that this means that the fibers of $\Gamma Q \to \Gamma P$ have cardinality strictly less than κ.)

4.1 Proposition. *This class of κ-small maps in Eff satisfies all the axioms (A1–7), (S1) and (S2) for small maps. Moreover, the separation axiom (S4) and the axiom of infinity (S5) are satisfied. If κ is inaccessible, the power-set axiom (S3) holds for the class of κ-small maps.*

Proof. The verification of the axioms for open maps (A1–7) is straightforward. By way of example, we verify closure under composition (A1), which is probably the least obvious case. For this, suppose $f : Y \to X$ and $g : Z \to Y$ are small, as witnessed by a diagram (2) for f and a similar diagram for g,

$$
\begin{array}{ccc}
S & \longrightarrow & Z \\
\downarrow & & \downarrow \\
R & \longrightarrow & Y.
\end{array}
$$

First construct the pullback $Q \times_Y R$. Since the projection $Q \times_Y R \to Q$ is epi, as a pullback of the epi $R \to Y$, and since Q is projective, this projection has a section $(1, \sigma) : Q \to Q \times_Y R$. For this map $\sigma : Q \to R$, form the pullback $Q \times_R S$. Then

$$
\begin{array}{ccc}
Q \times_R S & \longrightarrow & Z \\
\downarrow & & \downarrow \\
P & \longrightarrow & X
\end{array}
$$

is readily seen to be a quasi-pullback. Moreover, since R is projective, hence separated, the pullback $Q \times_R S$ is a double-negation-closed subobject of $Q \times S$; hence $Q \times_R S$ is projective. Finally, since the functor $\Gamma : Eff \to Sets$ preserves pullbacks, the properties of κ-small maps in *Sets* immediately imply that $\Gamma(Q \times_R S) \to \Gamma(Q)$, and hence the composite $\Gamma(Q \times_R S) \to \Gamma(Q) \to \Gamma(P)$, are κ-small. This proves that the preceding diagram witnesses that the composition $f \circ g : Z \to X$ is a κ-small map in *Eff*.

The Exponentiability Axiom (S1) holds since *Eff* is an elementary topos. Postponing the proof of the Representability Axiom (S2), we observe that the Axiom of infinity (S5) holds, since the natural numbers object N of *Eff* is projective while $\Gamma(\mathsf{N})$ is the set of natural numbers.

For the Separation Axiom (S4), consider a monomorphism $Y \rightarrowtail X$, and let $P \twoheadrightarrow X$ be any projective cover of X. Then $S = P \times_X Y$ is a subobject of P, hence is separated. Form the canonical projective cover $\tilde{S} \twoheadrightarrow S$. Then the square

$$
\begin{array}{ccc}
\tilde{S} \longrightarrow P \times_X Y \longrightarrow Y \\
\downarrow \qquad\qquad\qquad\qquad\qquad \downarrow \\
P \xrightarrow{\qquad\qquad\qquad\qquad\qquad} X
\end{array}
$$

is a quasi-pullback. Furthermore, since $\tilde{S} \subseteq S \times \mathsf{N} \subseteq P \times \mathsf{N}$, the map $\Gamma(\tilde{S}) \to \Gamma(P)$ has countable fibers, hence is small.

Finally, for the representability axiom (S2) and the powerset axiom (S3), we use the following equivalent description of the κ-small maps in *Eff*, in terms of the internal logic of *Eff*. (In this description, $P(\Delta K)$ denotes the power object $\Omega^{\Delta(K)}$ in *Eff*.)

4.2 Lemma. *Let K be a fixed set of cardinality κ. A map $f : Y \to X$ is κ-small iff the sentence*

$$
\forall x \in X \; \exists Q, T \in P(\Delta K) \; [\exists \text{ epi } e : \; Q \to f^{-1}(x), \text{ and}
$$

$$
Q \subseteq T = \neg\neg T, \text{ and } \neg \; \exists \text{ epi } T \to \Delta(K)]
$$

is valid in Eff.

Proof. ($\Rightarrow$) Let $\varphi : Q \to P$ be the map in a square of the form (2), witnessing that $f : Y \to X$ is κ-small. Since $\Gamma(\varphi) : \Gamma(Q) \to \Gamma(P)$ is a κ-small map of sets, there exists an embedding of each fiber of $\Gamma(\varphi)$ into the set K, thus giving a mono $\Gamma(Q) \to \Gamma(P) \times K$ over $\Gamma(P)$. By exponential transposition, we thus obtain a map

$$
Q \to \Delta(\Gamma(P) \times K) \cong \Delta\Gamma(P) \times \Delta(K) \tag{3}
$$

over $\Delta\Gamma(P)$. This map is again mono. (Indeed, if S is any separated object of *Eff* and L is any set, then for a mono $\Gamma(S) \rightarrowtail L$ the transposed map $S \to \Delta(L)$ is again mono, since Δ preserves monos while the unit $S \to \Delta\Gamma(S)$ is mono.) Pulling back the mono (3) along the unit $P \to \Delta\Gamma(P)$ yields a mono $Q \rightarrowtail P \times \Delta(K)$ over P. This mono is a subobject of $\Delta(K)$ in *Eff*/P. Moreover, writing x for the point of X in *Eff*/P given by the map $P \to X$ in (2), the fact that (2) is a quasi-pullback gives an epi $Q \to f^{-1}(x)$ in *Eff*/P. Finally, let $T \rightarrowtail P \times \Delta(K)$ be the pullback of $T' = \Delta\Gamma(Q) \rightarrowtail \Delta\Gamma(P) \times \Delta(K)$ along $P \to \Delta\Gamma(P)$. Now in *Eff*/$\Delta\Gamma(P)$, the sentence

$$
\text{`` } \neg\neg T' = T' \text{ and } \neg\neg\exists \text{ epi } T' \to \Delta(K) \text{ ''}
$$

is easily seen to be valid, since $\Gamma(Q) \to \Gamma(P)$ has all its fibers of cardinality strictly below the cardinality of K. This sentence remains valid after pulling back to Eff/P, showing that

$$\text{`` } \neg\neg T = T \text{ and } \neg\neg\exists \text{ epi } T \to \Delta(K) \text{ ''}$$

is valid in Eff/P. This proves that

$$Eff/P \models \exists Q, T \in P(\Delta K)[\cdots],$$

as in the statement of the lemma, for the map $x : P \to X$. Since this map is epi, it follows that

$$Eff \models \forall x \in X \; \exists Q, T \in P(\Delta K)[\cdots],$$

as required.

($\Leftarrow$) For a map $f : Y \to X$, validity of the sentence in the statement of the lemma means that there exists a diagram

$$\begin{array}{ccc}
T & \longleftarrow Q \longrightarrow & Y \\
\downarrow & \downarrow & \downarrow \\
P \times \Delta(K) & \longrightarrow P \longrightarrow & X
\end{array} \tag{4}$$

where $T \subseteq P \times \Delta(K)$ is double-negation-closed, such that the right-hand square is a quasi-pullback while "$\neg\exists$ epi $T \to \Delta(K)$" holds in Eff/P. We may assume that P is projective. Then T is also projective, being a double-negation-closed subobject of the projective object $P \times \Delta(K)$. Furthermore, the fact that $Eff/P \models$ "$\neg \exists$ epi $T \to \Delta(K)$" readily implies that all the fibers of $\Gamma(T) \to \Gamma(P)$ have cardinality strictly below κ. To complete the proof, note that Q is separated, and let $\tilde{Q} \subseteq Q \times \mathbf{N}$ be its canonical projective cover. Then

$$\begin{array}{ccc}
\tilde{Q} & \longrightarrow & Y \\
\downarrow & & \downarrow \\
P & \longrightarrow & X
\end{array}$$

is a quasi-pullback, and $\Gamma(\tilde{Q}) \to \Gamma(P)$ is κ-small in $Sets$, since $\Gamma(\tilde{Q}) \to \Gamma(Q)$, $\Gamma(Q) \to \Gamma(T)$ and $\Gamma(T) \to \Gamma(P)$ all are. Thus (4) witnesses that $Y \to X$ is a κ-small map in Eff.

This proves the lemma.

The proof of Proposition 4.1 can now easily be completed. Indeed, by

Lemma 4.2, it follows that "the" universal small map $\pi : E \to U$ can be constructed using the internal logic of *Eff*, as follows. Let

$$U = \{(S, R, T) \mid S \subseteq T = \neg\neg T, \ R \text{ is an equivalence}$$

$$\text{relation on } S, \text{ and } \neg\exists \text{ epi } T \to \Delta(K)\},$$

and define $E \to U$ as the internal sum

$$E = \sum_{(S,R,T)\in U} S/R.$$

Universality of $E \to U$ is immediate from Lemma 4.2.

Finally, for the power-set axiom (S3), note that since the separation axiom (S4) holds, we have for any small object X in *Eff* that $P_s(X) = P(X) = \Omega^X$. For an inaccessible κ, this exponential Ω^X is clearly again small. Similarly, one readily verifies that $P(X \to B)$ is a κ-small object in *Eff/B* whenever $X \to B$ is, for an inaccessible κ.

This completes the proof of Proposition 4.1.

§5 Choice maps

The previous example is in fact an illustration of a more general phenomenon. Recall that an object A in a topos $\mathcal{C}$ is said to be *internally projective* if $(-)^A$ preserves epimorphisms. There is an obvious equivalent description, which does not use exponentials and makes sense in any pretopos: A is internally projective iff for any epimorphism $Y \twoheadrightarrow X$ and any arrow $T \times A \to X$ there exist an epi $T' \twoheadrightarrow T$ and a map $T' \times A \to Y$ such that the square

$$\begin{array}{ccc} T' \times A & \longrightarrow & Y \\ \downarrow & & \downarrow \\ T \times A & \longrightarrow & X \end{array}$$

commutes. In terms of the logic of $\mathcal{C}$, an object A is internally projective iff the internal axiom of choice holds for quantifiers of the form

$$\forall a \in A \ \exists x \in X(\cdots)$$

where X is any other object of $\mathcal{C}$. Thus we say that A is a *choice object* if A is internally projective. More generally, a map $B \to A$ is said to be a *choice map* if $B \to A$ is internally projective as an object in $\mathcal{C}/A$. In terms of the logic of $\mathcal{C}$, this means that the axiom of choice along the map $B \to A$ is valid.

We note the following simple closure properties of the class of choice maps in a pretopos $\mathcal{C}$:

5.1 Proposition. *In any pretopos $\mathcal{C}$, the class of all choice maps satisfies the axioms (A1-5) for open maps.*

Now let $\mathcal{A}$ be any class of choice maps in the pretopos $\mathcal{C}$, which contains all isomorphisms, is closed under composition, pullbacks and sums, and contains $0 \to 1$ as well as $1 + 1 \to 1$ (i.e., $\mathcal{A}$ satisfies axioms (A1,2,4,5)). Define a map $Y \to X$ to be $\mathcal{A}$-small if there exists a quasi-pullback

$$
\begin{array}{ccc}
B & \longrightarrow & Y \\
\downarrow & & \downarrow \\
A & \longrightarrow & X
\end{array}
\tag{1}
$$

where $B \to A$ belongs to $\mathcal{A}$. Denote this class of $\mathcal{A}$-small maps by $\mathcal{S}(\mathcal{A})$.

5.2 Proposition. *The class $\mathcal{S}(\mathcal{A})$ satisfies all the axioms (A1–7) for open maps.*

The proof of this proposition is routine, and omitted.

Now suppose $\mathcal{C}$ is an elementary topos, and assume that the class of choice maps $\mathcal{A}$ is representable; i.e., there exists a "universal" map $\pi : E \to U$ in $\mathcal{A}$ such that for every other map $f : Y \to X$ in $\mathcal{A}$ there exists a double pullback of the form of diagram (4) in Chapter I, §1. It follows that the class of $\mathcal{A}$-small maps $\mathcal{S}(\mathcal{A})$ is representable as well. Indeed, let

$$
V = \{(u, R) \mid u \in U, \ R \text{ an equivalence relation on } E_u\}.
$$

This object V can be constructed in $\mathcal{C}$, as a subobject of the power-object $P(E \times_U E \to U)$ in the topos $\mathcal{C}/U$. Then $V \times_U E \to V$ is the universal map in $\mathcal{A}$ equipped with a fiberwise equivalence relation R, and the universal $\mathcal{A}$-small map is constructed as the quotient $F = (V \times_U E)/R \to V$.

Thus Proposition 5.2 can be extended as follows.

5.3 Proposition. *Let $\mathcal{A}$ be a class of choice maps in a topos $\mathcal{E}$, satisfying axioms (A1), (A2), (A4) and (A5) for open maps, as well as the Representability Axiom (S2). Then the class $\mathcal{S}(\mathcal{A})$ of $\mathcal{A}$-small maps satisfies* *all the axioms for small maps.*

For an example, define a map $Y \to X$ in the effective topos *Eff* to be *quasi-modest* if there exists a diagram of the form

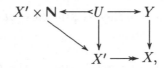

where the right-hand square is a quasi-pullback and U is a double-negation-closed subobject of $X' \times \mathsf{N}$ in the topos Eff/X' (here N denotes the natural numbers object of *Eff*). So a map $Y \to X$ is quasi-modest if its fibers are subcountable, in the strong sense of being quotients of double-negation-closed subsets of N.

These quasi-modest maps are closely related to the category of what are now commonly called the modest sets . In fact, if we call a map $Y \to X$ modest if, in addition to being quasi-modest, its diagonal $Y \to Y \times_X Y$ is double-negation-closed, then the modest sets are exactly the objects Y for which $Y \to 1$ is a modest map.

5.4 Proposition. *The class of quasi-modest maps satisfies all the axioms for small maps, as well as the additional axioms (S4) for separation and (S5) for infinity.*

Proof. Consider in the effective topos *Eff* the class $\mathcal{A}$ consisting of all maps $B \to A$ which are double-negation-closed subobjects of N in Eff/A. Each such $B \to A$ is a choice map. Indeed, if $P \twoheadrightarrow A$ is any projective cover of A, then $B \times_A P$ is a double-negation-closed subobject of $\mathsf{N} \times P$. Since N and P are both projective, so is $B \times_A P$. It follows that $B \to A$ is internally projective, i.e., a choice map. This class $\mathcal{A}$ is evidently closed under pullbacks and sums, and contains $0 \to 1$ as well as $1 + 1 \to 1$. Furthermore, using the isomorphism $\mathsf{N} \times \mathsf{N} \cong \mathsf{N}$, one readily verifies that $\mathcal{A}$ is closed under composition. Finally, $\mathcal{A}$ is clearly representable: a universal member of $\mathcal{A}$ obtained by pulling back the membership relation $\in_\mathsf{N} \rightarrowtail \mathsf{N} \times P(\mathsf{N}) \to P(\mathsf{N})$ along the inclusion $P_{\neg\neg}(\mathsf{N}) \rightarrowtail P(\mathsf{N})$ given by the double-negation-closed subsets of N. By Proposition 5.3, the associated class $\mathcal{S}(\mathcal{A})$ satisfies the axioms for small maps. It also clearly satisfies the Axiom of Infinity (S5), since $\mathsf{N} \to 1$ belongs to $\mathcal{A}$. Finally, to prove that $\mathcal{S}(\mathcal{A})$ satisfies the Separation Axiom, suppose $Y \to X$ is mono. Let $T \twoheadrightarrow X$ be a separated cover of X, and write $S = Y \times_X T \rightarrowtail T$. Then S is also separated, so S has a canonical projective cover $\tilde{S} \twoheadrightarrow S$, with $\tilde{S} \subseteq S \times \mathbf{N}$ over S. The composite inclusion $\tilde{S} \subseteq S \times \mathbf{N} \subseteq T \times \mathbf{N}$ represents a double-negation-closed subobject of $T \times \mathbf{N}$. Therefore $Y \to X$ belongs to $\mathcal{S}(\mathcal{A})$.

This completes the proof.

Appendix A. Monads and algebras with successor

The purpose of this appendix is to present an abstract version of the proof of Theorem II.1.2. For the basic definitions concerning monads and their algebras, we refer the reader to Mac Lane(1971), Chapter VI.

Let $\mathcal{C}$ be an arbitrary category, and let $\mathbf{P} = (P, \sigma, \mu)$ be a monad on $\mathcal{C}$. Let $\mathbf{P}\text{-Alg}(\mathcal{C})$ be the category of $\mathbf{P}$-algebras $(X, a : P(X) \to X)$. Recall that each object $X \in \mathcal{C}$ generates a free $\mathbf{P}$-algebra

$$(P(X), \mu_X : P^2(X) \to P(X)). \tag{1}$$

A $\mathbf{P}$-algebra with a successor operation – briefly, a *successor algebra* – is a $\mathbf{P}$-algebra (X, a) equipped with a map $s : X \to X$. With algebra homomorphisms which respect the successor operation, these successor algebras form an obvious category

$$\mathbf{P}\text{-Alg succ}(\mathcal{C}).$$

A.1 Theorem. *Suppose* (V, a, s) *is an initial object in the category* $\mathbf{P}\text{-Alg succ}(\mathcal{C})$ *of successor algebras in* $\mathcal{C}$. *Then the map*

$$r : P(V) \to V \ , \quad r = a \circ P(s)$$

is an isomorphism.

Proof. The proof will be an almost verbatim transcription of the proof of Theorem II.1.2, and, to facilitate the comparison, we will use the same notation as much as possible.

First, the algebra $(P(V), \mu)$ is the free algebra on V, so the successor map $s : V \to V$ can be extended uniquely along the unit $\sigma : V \to P(V)$ to an algebra homomorphism $r : (P(V), \mu) \to (V, a)$:

$$
\begin{array}{ccc}
P(V) & \overset{r}{\dashrightarrow} & V \\
{\scriptstyle \sigma}\big\uparrow & \nearrow {\scriptstyle s,} & \\
V & &
\end{array}
\qquad
\begin{array}{l}
r \circ \sigma = s, \\[4pt]
r \circ \mu = a \circ P(r).
\end{array}
$$

Observe that the successor $s : V \to V$ induces a homomorphism of free algebras $P(s) : (P(V), \mu) \to (P(V), \mu)$, while one can view the algebra structure $a : P(V) \to V$ as a homomorphism $a : (P(V), \mu) \to (V, a)$. This gives a composite homomorphism $a \circ P(s)$, which satisfies the defining identity for r:

$$a \circ P(s) \circ \sigma = a \circ \sigma \circ s = s.$$

Thus, by uniqueness of r,

$$r = a \circ P(s),$$

as in the statement of the theorem.

Now define a successor s' on the free algebra $(P(V), \mu)$ by

$$s' = \sigma \circ r : P(V) \to P(V).$$

Since (V, a, s) is the initial successor algebra, there exists a unique map $i : V \to P(V)$ which is an algebra homomorphism,

$$\mu \circ P(i) = i \circ a,$$

and preserves the successor,

$$i \circ s = s' \circ i.$$

We will show that i and r are mutually inverse isomorphisms. For $r \circ i = 1_V$, it suffices (by initiality of V) to prove that the algebra map r preserves the successor. This is indeed the case, since

$$r \circ s' = r \circ \sigma \circ r = s \circ r.$$

For $i \circ r = 1_{P(V)}$, it suffices by freeness of $(P(V)), \mu)$ to prove that $i \circ r \circ \sigma = \sigma$. But

$$\begin{aligned}
i \circ r \circ \sigma &= i \circ s \\
&= s' \circ i \\
&= \sigma \circ r \circ i \\
&= \sigma,
\end{aligned}$$

the last since we have proved already that $r \circ i = 1_V$.

This proves the theorem.

There is a similar abstract version of Theorem II.1.5. Let A be an object of $\mathcal{C}$, and suppose the category $\mathcal{C}$ has finite sums. (In fact we only need sums $A + (-)$.) For maps $f : A \to Y$ and $g : B \to Y$ in $\mathcal{C}$, we write $[f, g]$ for the induced map $A + B \to Y$. The *free successor algebra generated by* A is a successor algebra $(V(A), a, s)$ with a map $\eta : A \to V(A)$, such that for any other successor algebra (W, b, t), any map $g : A \to W$ can be uniquely extended along η to a homomorphism of successor algebras $\tilde{g} : (V(A), a, s) \to (W, b, t)$.

A.2 Theorem. *Suppose* $(V(A), a, s)$ *is the free successor algebra generated by* A *(via the map* $\eta : A \to V(A)$*). Then the map*

$$r : P(A + V(A)) \to V(A) \ , \ r = a \circ P([\eta, s])$$

is an isomorphism.

Using some monad theory, this theorem can be seen to be a consequence of Theorem A.1, see Remark A.3 below. We first give a direct proof, analogous to the proof of Theorem II.1.5.

Proof. Since $(P(A + V(A)), \mu)$ is the free algebra on $A + V(A)$ (as in (1) above), the map $[\eta, s]$ can be extended uniquely to an algebra homomorphism r, as in

$$P(A + V(A)) \overset{r}{-\,-\,\blacktriangleright} V(A)$$

$$\sigma \uparrow \qquad \nearrow$$
$$\qquad \overset{[\eta, s],}{}$$
$$A + V(A)$$

$$r \circ \sigma = [\eta, s],$$
$$r \circ \mu = a \circ P(r).$$

Since the composite homomorphism of algebras

$$(P(A + V(A)), \mu) \overset{P([\eta, s])}{\longrightarrow} (P(V(A)), \mu) \overset{a}{\to} (V(A), a)$$

also satisfies the defining identity for r, i.e.

$$a \circ P([\eta, s]) \circ \sigma = \quad a \circ \sigma \circ [\eta, s]$$
$$= \quad [\eta, s],$$

it follows that

$$r = a \circ P([\eta, s]),$$

as in the statement of the theorem.

Now define a map

$$\eta' : A \to P(A + V(A)) \ , \ \eta' = \sigma \circ c_1,$$

where $c_1 : A \to A + V(A)$ is the first coproduct inclusion, and $\sigma = \sigma_{(A+V(A))}$ is the unit of the monad P. Also define a successor

$$s' : P(A + V(A)) \to P(A + V(A)), \quad s' = \sigma \circ c_2 \circ r$$

(with $c_2 : V(A) \to A + V(A)$ the second coproduct inclusion). This makes $(P(A+V(A)), \mu, s')$ into a successor algebra. Since $V(A)$ is the free successor algebra on A, there is a unique map

$$i : V(A) \to P(A + V(A))$$

which is an algebra homomorphism,

$$\mu \circ P(i) = i \circ a,$$

preserves the sucessor,

$$i \circ s = s' \circ i,$$

and extends η',

$$i \circ \eta = \eta'.$$

As in the previous proof, r and i are mutually inverse isomorphisms. Indeed, r is in fact a homomorphism of successor algebras, since

$$
\begin{aligned}
r \circ s' &= r \circ \sigma \circ c_2 \circ r \\
&= [\eta, s] \circ c_2 \circ r \\
&= s \circ r.
\end{aligned}
$$

Furthermore,

$$
\begin{aligned}
r \circ \eta' &= r \circ \sigma \circ c_1 \\
&= [\eta, s] \circ c_1 \\
&= \eta,
\end{aligned}
$$

thus r "preserves" the generators. It follows that $r \circ i$ is a homomorphism of successor algebras such that $r \circ i \circ \eta = r \circ \eta' = \eta$. Hence $r \circ i$ must be the identity map, since $V(A)$ is freely generated by A.

To show that the other composite $i \circ r$ is the identity map, it suffices to prove that $i \circ r \circ \sigma = \sigma : A + V(A) \to P(A + V(A))$, because $(P(A + V(A)), \mu)$ is the free algebra on $A + V(A)$. But

$$
\begin{aligned}
i \circ r \circ \sigma &= i \circ [\eta, s] \\
&= [i \circ \eta, \; i \circ s] \\
&= [\eta', \; s' \circ i] \\
&= [\eta', \; \sigma \circ c_2 \circ r \circ i] \\
&= [\sigma \circ c_1, \; \sigma \circ c_2] \\
&= \sigma.
\end{aligned}
$$

This shows that $i \circ r = id$, and completes the proof of the theorem.

A.3 Remark. Using distributive laws for the composition of monads (Beck(1969)), Theorem A.2 can be seen as a special case of Theorem A.1. Indeed, note that for an object A of $\mathcal{C}$, there is a monad

$$\mathsf{A} = (A + (-), c_2, \nabla),$$

with as unit the second coproduct inclusion $c_2 : X \to A + X$, and as multiplication the codiagonal $\nabla_X : A + A + X \to A + X$. An A-algebra structure on an object X is the same thing as a map $h : A \to X$. Or in other words, the category $(\mathsf{A}\text{-Alg})$ of A-algebras is (equivalent to) the slice category A.

If $\mathsf{P} = (P, \sigma, \mu)$ is any other monad, there is an obvious natural map

$$\delta_X : A + P(X) \to P(A + X),$$

which satisfies the equations for a distributive law given in Beck(1969). Thus there is a composite monad $\mathsf{P} \circ \mathsf{A}$ with as underlying functor $X \mapsto P(A + X)$. And a $(\mathsf{P} \circ \mathsf{A})$-algebra structure on an object X is given by a map $h : A \to X$ and a P-algebra structure $a : P(X) \to X$. The initial $(\mathsf{P} \circ \mathsf{A})$-algebra (with successor) is the same thing as the free P-algebra (with successor) generated by A. Thus Theorem A.1, applied to the composite monad $\mathsf{P} \circ \mathsf{A}$, yields the isomorphism of Theorem A.2.

It was pointed out to us by J. Bénabou, and, independently, by M. Jidbladze, that our Theorem A.1 is related to a similar result of Lambek(1970). To explain this relation, consider any endofunctor $P : \mathcal{C} \to \mathcal{C}$ (not necessarily the underlying functor of a monad), and define a P-algebra to be an object X with a map $a : P(X) \to X$. With the evident morphisms, these P-algebras form a category. If P is part of a monad $\mathsf{P} = (P, \sigma, \mu)$, we will call such P-algebras weak P-algebras , to emphasize the distinction between such and the algebras for the monad P.

We recall Lambek's result:

A.4 Proposition.(Lambek). *Let $P : \mathcal{C} \to \mathcal{C}$ be an endofunctor. If $(V, h : P(V) \to V)$ is an initial (weak) P-algebra, then the algebra map h is an isomorphism.*

Proof. For the weak algebra $(P(V), P(h) : P^2(V) \to P(V))$, the initiality of (V, h) gives a unique homomorphism $i : (V, h) \to (P(V), P(h))$; in particular

$$P(h) \circ P(i) = i \circ h.$$

Also, h is a homomorphism $h : (P(V), P(h)) \rightarrow (V, h)$. Since (V, h) is initial, the composite homomorphism $h \circ i$ must be the identity. But then $i \circ h = P(h) \circ P(i) = P(h \circ i) = P(1) = 1$. Thus i and h are mutually inverse isomorphisms.

The following result now relates Lambek's isomorphism to our isomorphism in Theorem A.1:

A.5 Theorem. (Bénabou, Jidbladze). *Let* $\mathbf{P} = (P, \sigma, \mu)$ *be a monad on the category* $\mathcal{C}$.

(i) *If* (V, a, s) *is an initial successor algebra then* $(V, a \circ P(s))$ *is an initial weak P-algebra.*

(ii) *If* (V, h) *is an initial weak P-algebra (so that h is iso by Proposition A.4) then* $(V, h\mu P(h^{-1}), h \circ \sigma)$ *is an initial successor algebra.*

(Note that these two constructions are inverse to each other.)

Proof. ($\Rightarrow$) Assume (V, a, s) is an initial successor algebra. To show that $(V, a \circ P(s))$ is an initial weak algebra, let $(X, b : P(X) \rightarrow X)$ be any other weak algebra. We will show that there is a unique homomorphism $V \rightarrow X$ of weak algebras. First, the free **P**-algebra $(P(X), \mu)$ on X comes equipped with a successor $\sigma \circ b : P(X) \rightarrow X \rightarrow P(X)$. Since (V, a, s) is initial, there is a unique homomorphism of successor algebras $(V, a, s) \rightarrow (P(X), \mu, \sigma b)$, i.e. a map

$$\varphi : V \rightarrow P(X)$$

such that

$$\mu \circ P(\varphi) = \varphi \circ a, \tag{2}$$

$$\varphi \circ s = \sigma \circ b \circ \varphi. \tag{3}$$

Then $b \circ \varphi : V \rightarrow X$ is a homomorphism of weak algebras, since

$$
\begin{aligned}
b \circ \varphi \circ a \circ P(s) &= b \circ \mu \circ P(\varphi) \circ P(s) && \text{(by (2))} \\
&= b \circ \mu \circ P(\varphi s) \\
&= b \circ \mu \circ P(\sigma b \varphi) && \text{(by (3))} \\
&= b \circ (\mu \circ P(\sigma)) \circ P(b\varphi) \\
&= b \circ P(b\varphi) && \text{(by the unit law for } \mathbf{P}).
\end{aligned}
$$

To see that $b\varphi : V \to X$ is the unique homomorphism of weak algebras, suppose $\psi : V \to X$ is any other such homomorphism, such that

$$\psi \circ a \circ P(s) = b \circ P(\psi). \tag{4}$$

Then for $s' = \sigma \circ r = \sigma \circ a \circ P(s)$ as in the proof of Theorem A.1, the map $P(\psi) : P(V) \to P(X)$ is a homomorphism of successor algebras $(P(V), \mu, s')$ $\to (P(X), \mu, \sigma b)$; indeed,

$$
\begin{aligned}
P(\psi) \circ s' &= P(\psi) \circ \sigma \circ a \circ P(s) \\
&= \sigma \circ \psi \circ a \circ P(s) \\
&= \sigma \circ b \circ P(\psi) \qquad \text{(by (4))}.
\end{aligned}
$$

Thus the composition of $P(\psi)$ with the unique homomorphism $i : (V, a, s) \to$ $(P(V), \mu, s')$ (already used in the proof of Theorem A.1) must be the unique homomorphism φ above, i.e., $\varphi = P(\psi) \circ i$. But then

$$
\begin{aligned}
\varphi &= \psi \circ r \circ i \\
&= b \circ P(\psi) \circ i \quad \text{(by (4), since } r = a \circ P(s)) \\
&= b \circ \varphi.
\end{aligned}
$$

This shows that $b \circ \varphi$ is unique, as required.

($\Leftarrow$) Suppose (V, h) is the initial weak algebra. Write

$$a := h \circ \mu \circ P(h^{-1}) \, , \quad s := h \circ \sigma.$$

To show that (V, a, s), thus defined, is the initial successor algebra, consider any other such algebra $(X, b : P(X) \to X, t : X \to X)$. Then $(X, b \circ P(t))$ is a weak algebra, so there is a unique homomorphism of weak algebras $\varphi : (V, h) \to (X, b \circ P(t))$; thus

$$b \circ P(t) \circ P(\varphi) = \varphi \circ h. \tag{5}$$

But then φ is also a homomorphism of successor algebras. Indeed, φ is an algebra homomorphism, since

$$
\begin{aligned}
\varphi \circ a &= \varphi \circ h \circ \mu \circ P(h^{-1}) \\
&= b \circ P(t) \circ P(\varphi) \circ \mu \circ P(h^{-1}) \qquad \text{(by (5))} \\
&= b \circ \mu \circ P^2(t\varphi) \circ P(h^{-1}) \qquad \text{(naturality of } \mu) \\
&= b \circ P(b) \circ P^2(t\varphi) \circ P(h^{-1}) \qquad \text{(associativity of } (X, b)) \\
&= b \circ P(b \circ P(t) \circ P(\varphi)) \circ P(h^{-1}) \\
&= b \circ P(\varphi h) \circ P(h^{-1}) \qquad \text{(by (5))} \\
&= b \circ P(\varphi),
\end{aligned}
$$

and φ preserves the successor, since

$$
\begin{aligned}
\varphi \circ s &= \varphi \circ h \circ \sigma & \text{(def. of } s) \\
&= b \circ P(t) \circ P(\varphi) \circ \sigma & \text{(by (5))} \\
&= b \circ \sigma \circ t \circ \varphi & \text{(naturality of } \sigma) \\
&= t \circ \varphi & \text{(unit law for } (X, b)).
\end{aligned}
$$

It remains to be shown that φ is the unique such homomorphism. To this end, suppose $\psi : V \to X$ is another homomorphism of successor algebras, i.e.

$$
\psi \circ a = b \circ P(\psi) \,, \ \psi \circ s = t \circ \psi. \tag{6}
$$

Then ψ is also a homomorphism of weak algebras, since

$$
\begin{aligned}
\psi \circ h &= \psi \circ h \circ \mu \circ P(\sigma) & \text{(unit law for } \mathbf{P}) \\
&= \psi \circ h \circ \mu \circ P(h^{-1}) \circ P(h) \circ P(\sigma) \\
&= \psi \circ a \circ P(s) & \text{(def. of } a \text{ and } s) \\
&= b \circ P(\psi) \circ P(s) & \text{(by (6))} \\
&= b \circ P(t) \circ P(\psi) & \text{(by (6))}.
\end{aligned}
$$

Since $\varphi : (V, h) \to (X, b \circ P(t))$ was defined to be the unique homomorphism of weak algebras, we must have $\psi = \varphi$.

This proves the theorem.

Appendix B. Heyting pretopoi

For the convenience of the reader, we recall in this appendix the definitions of the relevant properties of the ambient category, usually denoted $\mathcal{C}$ in the preceding chapters.

A category $\mathcal{C}$ is said to be a *pretopos* if it has the following properties P1–4.

(P1) $\mathcal{C}$ has pullbacks and a terminal object (hence all finite limits).

(P2) $\mathcal{C}$ has finite sums, and these are disjoint and stable under pullback.

Thus in particular $\mathcal{C}$ has an initial object 0 (the empty sum); disjointness means that for a finite sum $Y = Y_1 + \cdots + Y_n$, the pullback $Y_i \times_X Y_j$ is isomorphic to 0 whenever $i \neq j$, and stability means that for any family $\{f_i : Y_i \to X \mid i = 1, \cdots, n\}$ (any $n \geq 0$) and any arrow $X' \to X$, the canonical map $\sum(X' \times_X Y_i) \to X' \times_X \sum Y_i$ is an isomorphism. For the other defining properties of a pretopos, say that a diagram $R \rightrightarrows X \to Y$ is *exact* if $X \to Y$ is the coequalizer of $R \rightrightarrows X$ while $R \rightrightarrows X$ is the kernel pair of $X \to Y$ (i.e., $R \cong X \times_Y X$ as subobjects of $X \times X$).

(P3) For any equivalence relation $R \rightrightarrows X$ there exists some arrow $X \to Y$ for which $R \rightrightarrows X \to Y$ is exact.
And for any epimorphism $X \to Y$ there exists $R \rightrightarrows X$ for which $R \rightrightarrows X \to Y$ is exact.

(P4) If $R \rightrightarrows X \to Y$ is exact, then for any arrow $Z \to Y$ the diagram $Z \times_Y R \rightrightarrows Z \times_Y X \to Z \times_Y Y = Z$ is again exact.

Pretoposes are categories with the structure necessary for interpreting coherent logic ; for more on pretoposes, see e.g. Artin et.al.(1972), Johnstone(1977) or Makkai-Reyes(1977).

To interpret (intuitionistic) first order logic in $\mathcal{C}$, some additional structure is needed. For each object X of $\mathcal{C}$, write $Sub_{\mathcal{C}}(X)$ for the poset of subobjects of X. The axioms for a pretopos imply that each such poset

109

$Sub_{\mathcal{C}}(X)$ is a distributive lattice. Furthermore, each arrow $f : X \to Y$ induces a lattice homomorphism by pullback,

$$f^{-1} : Sub_{\mathcal{C}}(Y) \to Sub_{\mathcal{C}}(X).$$

This pullback map has a left adjoint

$$\exists_f : Sub_{\mathcal{C}}(X) \to Sub_{\mathcal{C}}(Y),$$

which sends a subobject $U \rightarrowtail X$ to the image of the composite $U \rightarrowtail X \to Y$. (The pretopos axioms imply that such images exist.) The category $\mathcal{C}$ is said to be a *Heyting pretopos* if each pullback map f^{-1} also has a right adjoint,

$$\forall_f : Sub_{\mathcal{C}}(X) \to Sub_{\mathcal{C}}(Y).$$

Thus, by the definition of adjoints, for $U \subseteq X$ and $V \subseteq Y$ one has $f^{-1}(V) \subseteq U$ iff $V \subseteq \forall_f(U)$. This operation $\forall_f$ is called universal quantification *along* f. Using these left adjoints, one can define an implication operator on each lattice $Sub_{\mathcal{C}}(X)$, making it into a Heyting algebra: for $U, V \in Sub_{\mathcal{C}}(X)$, write i for the inclusion $U \rightarrowtail X$, and define

$$(U \Rightarrow V) \;=\; \forall_i(U \cap V).$$

In Chapter III, we considered pretoposes which possess a *subobject classifier*. This is an object Ω, equipped with a "universal" subobject $t : 1 \to \Omega$; this means that for any other monomorphism $U \rightarrowtail X$, there is a unique arrow $c_U : U \to \Omega$ so that $U = c_U^{-1}(t)$, as in the pullback

$$
\begin{array}{ccc}
U & \longrightarrow & 1 \\
\downarrow & & \downarrow {\scriptstyle t} \\
X & \xrightarrow[c_U]{} & \Omega.
\end{array}
$$

Recall that an *elementary topos* is a pretopos $\mathcal{C}$ with a subobject classifier Ω, in which every object is exponentiable (cf. e.g. Johnstone(1977), Mac Lane-Moerdijk(1992)).

A pretopos $\mathcal{C}$ is said to be *Boolean* if every monomorphism $U \to X$ has a complement, i.e. a subobject $U^c \rightarrowtail X$ such that $X \cong U + U^c$ (by the maps $U \rightarrowtail X$ and $U^c \rightarrowtail X$). Thus in a Boolean pretopos, the subobject lattices $Sub_{\mathcal{C}}(X)$ are in fact all Boolean algebras. Every Boolean pretopos is Heyting (define $\forall_f(U)$ as $\exists_f(U^c)^c$), and has a subobject classifier (namely, $\Omega = 1 + 1$, with for $t : 1 \to \Omega$ one of the coproduct inclusions).

Finally, a *natural numbers object* in a pretopos $\mathcal{C}$ is an object $\mathbf{N}$, which is equipped with arrows $z : 1 \to \mathsf{N}$ (zero) and $s : \mathsf{N} \to \mathsf{N}$ (successor), and which is initial with this structure. Thus for any (parameter) object P and any arrows $f : P \to Y$ and $t : P \times Y \to Y$ in $\mathcal{C}$, there is a unique $\bar{f} : P \times \mathsf{N} \to Y$ for which the diagram

$$
\begin{array}{ccccc}
P \times 1 & \xrightarrow{id \times z} & P \times \mathsf{N} & \xrightarrow{id \times s} & P \times \mathsf{N} \\
\Big\| {\scriptstyle \iota} & & \Big\downarrow {\scriptstyle (\pi_1, \bar{f})} & & \Big\downarrow {\scriptstyle \bar{f}} \\
P & \xrightarrow{(id, f)} & P \times Y & \xrightarrow{\quad t \quad} & Y
\end{array}
$$

commutes.

(If $\mathcal{C}$ is cartesian closed, it suffices to require this for $P = 1$, which gives a perhaps more familiar version of the definition of a natural numbers object.)

Appendix C. Descent

We conclude this appendix with a brief discussion of effective descent in a pretopos $\mathcal{C}$. Let $f : Y \to X$ be a map in $\mathcal{C}$, and let $p : E \to Y$ be an object over Y. Descent data on $E \to Y$ (relative to f) are given by a map

$$\theta : Y \times_X E \to E$$

which makes the following three diagrams commute:

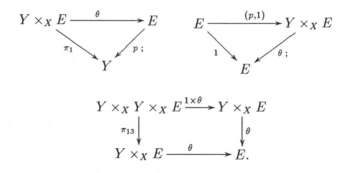

For sets, we can denote θ by a family of maps

$$\theta_{y_1,y_2} = \theta(y_2, -) : p^{-1}(y_1) \to p^{-1}(y_2)$$

and the commutativity of these diagrams corresponds to the equations

$$p\theta_{y_1,y_2}(e) = y_2,$$
$$\theta_{y,y}(e) = e,$$
$$\theta_{y_2,y_3}(\theta_{y_1,y_2}(e)) = \theta_{y_1,y_3}(e).$$

If $(E \xrightarrow{p} Y, \theta)$ and $(E' \xrightarrow{p'} Y, \theta')$ are two such objects over Y equipped with descent data, then one considers maps $u : E \to E'$ in $\mathcal{C}$ over Y (i.e., $p' \circ u = p$)

113

which respect descent data (i.e., $u \circ \theta = \theta' \circ (1 \times u)$). In this way, one obtains a category

$$Des(f)$$

of objects over Y equipped with descent data relative to f.

If $D \to X$ is any object over X, then the pullback $f^*(D) = (\pi_1 : Y \times_X D \to Y)$ has canonical descent data, namely $\theta = \pi_{13} : Y \times_X Y \times_X D \to Y \times_X D$. Thus, one has a functor

$$f^* : \mathcal{C}/X \to Des(f). \tag{1}$$

When this functor is an equivalence of categories, the map f is said to be an effective descent morphism (Grothendieck(1959)).

It follows readily from the axioms that, in a pretopos, every epimorphism is an effective descent map. Indeed, the condition that the functor f^* in (1) is full and faithful comes down to the fact that for any two objects D and D' over X, any arrow $u : Y \times_X D \to Y \times_X D'$ over Y which makes both left-hand squares in the diagram

$$\begin{array}{ccccc}
Y \times_X Y \times_X D & \underset{\pi_{13}}{\overset{\pi_{23}}{\rightrightarrows}} & Y \times_X D & \overset{\pi_2}{\longrightarrow} & D \\
\downarrow{\scriptstyle 1 \times u} & & \downarrow{\scriptstyle u} & & \downarrow{\scriptstyle v} \\
Y \times_X Y \times_X D' & \underset{\pi_{13}}{\overset{\pi_{23}}{\rightrightarrows}} & Y \times_X D' & \overset{\pi_2}{\longrightarrow} & D'
\end{array} \tag{2}$$

commute gives rise to a unique map $v : D \to D'$ (over X) making the whole diagram commute. But this follows immediately from the axioms for a pretopos. Indeed, since $f : Y \to X$ is assumed an epimorphism, the diagram

$$Y \times_X Y \rightrightarrows Y \to X$$

is exact, as is any pullback of it. In particular, both rows in (2) are exact, so that a unique v exists, as required.

To see that the functor f^* in (1) is essentially surjective, take any object $p : E \to Y$ equipped with descent data θ, and consider the associated equivalence relation $\theta, \pi_2 : Y \times_X E \rightrightarrows E$. This gives rise to a diagram with exact rows,

$$\begin{array}{ccccc}
Y \times_X E & \underset{\pi_2}{\overset{\theta}{\rightrightarrows}} & E & \overset{q}{\longrightarrow} & Q \\
\downarrow & & \downarrow{\scriptstyle p} & & \downarrow{\scriptstyle r} \\
Y \times_X Y & \underset{\pi_2}{\overset{\pi_1}{\rightrightarrows}} & Y & \underset{f}{\longrightarrow} & X
\end{array}$$

for a unique map r. In particular, one obtains a map $(p, q) : E \to Y \times_X Q \cong f^*(Q)$. On the other hand, exactness of the diagram

$$Y \times_X Y \times_X E \overset{1 \times \theta}{\underset{\pi_{13}}{\rightrightarrows}} Y \times_X E \overset{1 \times q}{\longrightarrow} Y \times_X Q,$$

together with the fact that $\theta \circ (1 \times \theta) = \theta \circ \pi_{13}$ by definition of descent data, yields a unique map $s : Y \times_X Q \to E$ such that $s \circ (1 \times q) = \theta$. It is easy to verify that s is a two-sided inverse for $(p, q) : E \to f^*(Q)$, and that under this isomorphism $E \cong f^*(Q)$, the given descent data θ on E correspond to the canonical descent data on $f^*(Q)$ (given by the definition of the functor f^* in (1)). This shows that f^* is essentially surjective, and proves that every epimorphism in a pretopos is an effective descent map.

Bibliography

P. **Aczel**, *Non-well-founded Sets*, CSLI Lecture Notes, Stanford, 1988.

M. **Artin**, A. **Grothendieck**, J.L. **Verdier**, *Théorie des Topos et Cohomologie étale des schémas* ("SGA4"), Springer Lecture Notes in Mathematics **269–270** (1972).

J. **Beck**, Distributive laws, in: B. Eckman(ed.), *Seminar on Triples and Categorical Homology Theory*, Springer Lectures Notes in Mathematics **80** (1969), 119–140.

J.L. **Bell**, *Boolean-valued Models and Independence Proofs in Set Theory*, Oxford University Press, Oxford, 1977.

A. **Blass**, A. **Scedrov**, Freyd's models for the independence of the axiom of choice, *Memoirs Amer. Math. Soc.* **404** (1989).

J.C. **Cole**, Categories of sets and models of set theory, in: J. Bell and A. Slomson (eds.), *Proc. B. Russell Memorial Logic Conference*, Leeds, 1973, 351–399.

M.P. **Fourman**, Sheaf models for set theory, *J. Pure and Applied Alg.* **19** (1980), 91–101.

P.J. **Freyd**, The axiom of choice, *J. Pure and Applied Alg.* **19** (1980), 103–125.

H. **Friedman**, Some applications of Kleene's method for intuitionistic systems, in: A.R.D. Mathias, H. Rogers (eds.), *Cambridge Summer School in Mathematical Logic*, Springer Lecture Notes in Mathematics. **337** (1973), 113–170.

117

W. Fulton, R. MacPherson, Categorical framework for the study of singular spaces, *Memoirs Amer. Math. Soc.* **243** (1981).

P. Gabriel, M. Zisman, *Calculus of Fractions and Homotopy Theory*, Springer-Verlag, Berlin, 1967.

R.J. Grayson, Heyting valued models for intuitionistic set theory, in: Springer Lecture Notes in Mathematics **753**(1977), 402–414.

A. Grothendieck, Technique de descente et théorèmes d'existence en géometrie algébrique, *Séminaire Bourbaki*(1959), exposé 190.

J.M.E. Hyland, The effective topos, in: *The Brouwer Centenary Symposium* (eds. A.S. Troelstra, D. van Dalen), North-Holland, Amsterdam, 1982, 165–216.

P.T. Johnstone, *Topos Theory*, Academic Press, New York 1977.

A. Joyal, I. Moerdijk, A completeness theorem for open maps, *Comptes Rendus Math., Soc. Royale du Canada* **12** (1990), 253–260.

A. Joyal, I. Moerdijk, A categorical theory of cumulative hierarchies of sets, *Comptes Rendus Math., Soc. Royale du Canada* **13** (1991), 55–58.

A. Joyal, I. Moerdijk, A completeness theorem for open maps, *Annals of Pure and Appl. Logic* **70** (1994), 51–86.

A. Joyal, M. Nielsen, G. Winskel, Bisimulation and open maps, in: *LICS 93 proceedings*, IEEE, 1993.

J. Lambek, Subequalizers, *Canadian Math. Bulletin* **13** (1970), 337–349.

D. McCarty, *Realizability and Recursive Mathematics*, Ph.D. Thesis, Carnegie-Mellon University, 1984.

S. Mac Lane, *Categories for the Working Mathematician*, Springer-Verlag, New York, 1971.

S. Mac Lane, I. Moerdijk, *Sheaves in Logic and Geometry*, Springer-Verlag, New York, 1992.

M. Makkai, G. Reyes, *First Order Categorical Logic*, Springer Lecture Notes in Mathematics **611** (1977).

W. Mitchell, Boolean topoi and the theory of sets, *J. Pure and Applied Alg.* **2** (1972), 261–274.

G. Osius, Categorical set theory: a characterization of the category of sets, *J. Pure and Applied Alg.* **4**(1974), 79–119.

E. Robinson, G. Rosolini, Colimit completions and the effective topos, *Journal of Symb. Logic* **55** (1990), 678–699.

P. Taylor, Intuitionistic sets and ordinals, preprint, 1994.

M. Tierney, Sheaf theory and the continuum hypothesis, in: F.W. Lawvere (ed.) *Toposes, Algebraic Geometry and Logic*, Springer Lecture Notes in Mathematics **274** (1972), 13–42.

Index